On the Wings of Genius:
A Chronicle of Modern Physics

Book I

*If I have seen farther, it is by standing
on the shoulders of giants.*

Isaac Newton (1676)

Dr. Andrew Worsley

*Universal Publishers
Boca Raton, Florida
USA • 2006*

On the Wings of Genius: A Chronicle of Modern Physics, Book I

Universal Publishers
Boca Raton, Florida • USA
2006

ISBN: 1-58112- 451-1

www.universal-publishers.com

About the Author:

Dr. Andrew Worsley (*nee:* Andrzej Wojciechowski):
Honorary Senior Lecturer, University of London.
Consultant, UHL, University of London.

About the Maths Editor:

Prof. Thomas Maibaum:
Professor of Mathematics,
Canada Research Chair in the Foundations of Software
Engineering.
Mc Masters University, Ontario, Canada.
Author of *The specification of Computer Programs*.

Front cover, Angel of Peace, by: John and Bridget Wren
Potter and Peter Ingram. Adapted from an original by
Gustave Doré (1832-83).

Back cover by: Gemma Worsley

Dedicated to:

The Truth and Beauty of Nature

All are but parts of one stupendous whole,
Whose body Nature is, and God the soul.

Alexander Pope (1688-1744)

Acknowledgements

This work is supported by the WWK Trust, a charitable trust, whose aims are to promote worldwide ecological, scientific and cultural human development, through the sharing of knowledge.

We gratefully acknowledge the assistance of Dr. Nikolaos Mavromatos, King's College London, for his scientific support and discussion.

My gratitude also to my wife, Carolyn, a total non-physicist, for her helpful comments and for making this book an enjoyable read.

Table of Contents

Foreword

Eppur, si muove - And yet, it moves.

Galileo Galilei (1633)

It has been said that a wise man does not share his knowledge, as by so doing, he allows everyone else to gain his knowledge and he is no longer seen as wise.

In truth, there is a difference between knowledge and wisdom, the two are not the same, but interlinked. The path to knowledge and the teaching of it is in part what bestows wisdom, as well as the knowledge itself. Furthermore, ignorance is not good, for the lack of knowledge, or even a little knowledge, is hazardous. If humankind had a more accurate and complete knowledge, that true knowledge would in itself endow a greater wisdom to use it properly. To this end, this book has been written to assist in knowing the beauty of modern physics, to help gain an insight into the elegance of the structure of the Universe and to give a far greater understanding of the aesthetic way in which Nature is designed.

Interestingly, Galileo published his knowledge of the workings of the solar system as a dialogue between three interlocutors, which had a popular appeal. Had Galileo not published in the way he did, his work on a heliocentric (sun centred) solar system may be unknown today. His work was heavily censured by the Church, which then believed that the Earth was the centre of the Universe and was immovable, and Galileo was forced to publicly recant

11

his works. Legend has it that after recanting his works, as he rose from kneeling, he quietly uttered the words – *and yet, it moves*. These words have captivated scientists and scholars for many years, as they epitomised a symbol of defiance and captivated the nobility of purpose, in adversity, in the search for truth and scientific beauty.

It is the manifest beauty and elegance of the Universe that *moves* the writing of this book. These chronicles of modern physics, are written on a historical basis and at the same time explain the aesthetics of quantum physics. The brilliant genius and the ironic twists and turns in the birth of modern physics are intriguing and ultimately point to a unified approach. Indeed, we introduce such a new unified and elegant view of the Universe, which corroborates modern physics and helps explain it. These new aspects of this work have also been published in a scientific fashion and I would guide the reader to the first two notes in the endnotes section.[1,2]

Great care has been taken in this book to maintain historical and scientific accuracy, and I stress that any new ideas do not challenge modern textbook physics but corroborate and help explain it.[3-5] Even some of the counter-intuitive propositions that are part of modern physics are explained in a logical fashion. Although quantum physics appears illogical, from the every day point of view, the new aspects to this work return us to a position where, not only is knowing the laws of Nature made much more enjoyable, but we *can* understand them.

Overall, this work brings an understanding of the beauty and symmetry of the Universe and the unified nature and the aesthetic elegance of its design.

Andrew Worsley, May 2005

Chapter 1

Introduction

I have been judged vehemently suspected of heresy, that is, of having held and believed that the sun is the centre of the universe and immovable, and that the earth is not the centre of the same, and that it does move.

Galileo Galilei (1633)

The subject of this present book is the history and explanation of modern physics and the introduction of a unified description of the Universe, which has truth, beauty [†] and symmetry.

Every so often science moves by large leaps, sometimes spurred by the realization that the commonly held beliefs of the day need revising. One such revolution was in the time of Galileo, at the end of the Renaissance. Another great change occurred at the turn of the twentieth century, when there was a revolution in modern scientific thinking. Yet another far greater turning point is about to occur.

Historically, the dawning of the modern era in science was initiated by Max Planck, in the early days of the twentieth century, by the fascinating discovery of a least energy quantity or quantum. This minimum energy unit was thenceforth called Planck's constant,

---

[†] Truth and beauty were the original names given to the two heaviest quarks.

and its discovery gave birth to an entirely new field of physics, known today as quantum physics.

Ironically, all this happened when scientists of the time had decided that all that could be known was known, apart from a few trivial details. This could not have been further from the truth.

The second irony of that discovery was that Planck was firmly embedded in old-fashioned Newtonian physics and initially had no idea how absolutely important his discovery of quantum physics was.

The third and most pertinent irony was that, had Planck at that time known and understood more about Einstein's later work on energy equivalence, almost the entirety of physics could have been ascertained by now. In actual fact, the knowledge we have today is small compared to the much broader and more elegant picture, which waits around the corner.

Today we hear of scientists saying again that they have discovered all that can be known apart from a few details. However, more recently there have been a few big surprises, particularly in Cosmology. It is now estimated to be some 13.7 billion years after the Big Bang, when the Universe exploded into being. Scientists had, by now, expected the expansion of the Universe, to be gradually slowing down. Somewhat intriguingly, the Universe is not only still expanding but is now doing so at an increasing rate.

This points us to new discoveries, which leads us to a new and beautifully unified approach to physics. In standard unified field theories, the ultimate aim is to unite the three known fundamental forces of nature. The first and strongest is the strong nuclear force, which effectively holds a nucleus of an atom together and gives nature that wonderful diversity of structure, which results from the production of differing atoms. The second is the electro-weak force, which gives us the electromagnetic spectrum; the most familiar example of this would be light. Such light gives us all the magnificent colours of the rainbow. The third is the gravitational force, which enables the formation of such exquisitely magnificent structures as the solar system and our own Milky Way galaxy, which rotates gracefully on its axis every 200 million years.

In this work we go far farther than merely unifying the forces of nature, but we also unify these forces with matter and space and time, so that the Universe becomes harmonious and beautiful. There is a subtle relationship between fundamental forces of nature and the elementary particles and space-time together. There are three forces of nature; these three forces in turn influence three types of particle or substance. Each of the particles is composed of multiples of the electric charge of 1/3, and the particles themselves are in three generations. All this is present in the three dimensions of real space. It is important that a truly comprehensive view of nature explains this threefold symmetry.

This points the way to a new scientific paradigm, and indeed we find that the answer to the mystery lies in what was discovered in the time of Planck. Not only does the answer to this mystery solve the problem of the expanding Universe and corroborate what is already know about quantum physics and the forces of nature, but it also leads us to a new unified, all-embracing and elegant description of the Universe.

The Sombrero Galaxy: in the centre of each galaxy lies a super-massive black hole, which is its gravitational dynamo. The glow comes from the trillions of stars, like our sun, which surround it in the galactic disc. Surrounding that is a belt of dark matter.

Chapter 2

Enormous Energy from Matter

Hitherto man had to live with the idea of death as an individual; from now on mankind will have to live with the idea of its death as a species.

Arthur Koestler

The important year was 1905, when Einstein wrote a number of seminal pieces of work. Perhaps the most well known was a paper on an equation for the equivalence of energy and matter, specifically $E = mc^2$. It was a very brief paper (only three pages long), and yet it has been the most often quoted and famous science equation of our times. Its impact was to have a huge effect on world history.

Originally in the 1905 paper, with the rather banal title: *Does the Inertia of a Body Depend on its Energy-Content?* Einstein wrote[8] :

"If a body gives off energy L in the form of radiation its mass diminishes by L/c²."

At that time, the term L was used for energy (probably from the German term *lebendige Kraft,* literally meaning "living" energy). The popular form of the equation, with the term *"E"* for energy did not emerge until 1912. Even then its full meaning was far from recognised.

Nor did Einstein's $E = mc^2$ make any obvious sense, where on Earth does it come from? Surely, there must be a more fundamental formula. Indeed, there is a more fundamental logical and beautiful energy equivalence formula, which we will elaborate in the next chapters. Suffice it to say that one should continue to question ones knowledge at the most fundamental levels, in order to gain new knowledge.

Not only was the true meaning of Einstein's formula not realised at the time, but the significance of the formula in terms of the potential for energy production was also not realised. It was not until 1939, when Einstein wrote to President Roosevelt, that the full theoretical potential of the energy equation $E = mc^2$ was beginning to be recognised. In his letter to Roosevelt, Einstein wrote:

"At stake, however, is more than abstract energy production. This new phenomenon would also lead to the construction of bombs, and it is conceivable...that extremely powerful bombs of a new type may thus be constructed."

Einstein's letter followed a practical breakthrough in atomic physics in 1939. It was then discovered that, provided you had sufficient Uranium, it was technically feasible to produce a very rapid radioactive chain reaction. It was then realised that a vast quantity of energy could be released from this process.

The only technical difficulty lay in the fact that one would require a critical mass for this chain reaction to be sustained. That critical mass in the case of ordinary Uranium was completely unwieldy – a ton (1,000 kg) of ordinary Uranium would be required. But there was a scientific trick to obviate this problem. The actual trick was to enrich the more reactive form of Uranium, then only about 20-25kg of this highly enriched Uranium would be required to sustain a chain reaction. In truth, the enrichment of this Uranium was technically and scientifically very difficult.

What was historically very fortunate was that in WWII, the Nazi Germans did not appear to realise this trick. The construction of a Nazi atom bomb would have otherwise been utterly devastating for entire the world. Indeed, there was a second breakthrough in atomic physics in 1941, which was even more dangerous. It was discovered that Plutonium could be used to make a nuclear bomb, and in this case, you only needed about 6kg of it. All you needed to do was construct a special nuclear reactor to produce the Plutonium, purify it, and then you could make an A-bomb.

It was ironic, and yet very fortunate for the entire world at that time, that a significant number of pre-eminent physicists in Europe were actually Jewish – and the lucky ones had fled Germany or Europe to go to the US or elsewhere. One such physicist was a Danish physicist called Niels Bohr. Historically, we know that in 1941, when Nazi Germany was at the height of its power, Bohr was visited in Copenhagen by his protégé and good friend, Werner Heisenberg. What we do not know, is what actually happened during that meeting. But *allegedly*, Heisenberg made Bohr an offer to build a Nazi nuclear bomb. We do know that, after a relatively short stay, Bohr announced that Heisenberg was leaving. Fortunately, it appears that Niel's Bohr had refused to build the Nazi nuclear bomb. In fairness to Heisenberg, he did report later to Hitler that it would be very difficult to build a nuclear bomb; as a result of that, an attempt to build a Nazi atomic bomb was dropped. As for Niels Bohr, he fled Copenhagen in 1941 to the US, just in the nick of time.

Similarly, Albert Einstein, the originator of the formula $E=mc^2$, had the foresight to flee Germany in 1933. Even at that time, however, he had not realised the actual power of his own equation. It was not until 1939, when Einstein wrote to President Roosevelt in an effort to beat Nazi Germany to the possibility of this new devastating technology, that its meaning became more widely apparent.

Indeed, the full meaning of the equation in terms of energy equivalence did not become clear until 1945, after the nuclear age began. When the first atomic bomb test was exploded, it was then clear that the technology was awesome. So much so, that it caused J. Robert Oppenheimer, who witnessed the test, to recall the words from the ancient Hindu script, *Baghavad Gita*:

"Now I am become death, the destroyer of worlds"

On august the 6th 1945, a B29 bomber called Enola Gay, dropped Little Boy on the city of Hiroshima. From that moment on it was clear that the world would no longer be the same.

What had perhaps caused it to change most was that little equation, $E = mc^2$. The equation looks straightforward – energy is equivalent to mass times the speed of light, squared. The equation meant that if a mass of m kilograms is lost in splitting an atom, such as a Uranium atom, and the resultant parts weigh less than the original, the difference in the mass is converted into energy. This enables the potential release of vast quantities of energy.

This results from the magnitude of the term for the speed of light. The speed of light itself is 3×10^8 meters per second, equivalent to 186,000 miles per second. So, if you square the speed of light in meters per second, you get an enormous number of 9×10^{16} (90,000,000,000,000,000). So for every kilogram of matter that is converted to energy, you get an enormous yield of energy of 9×10^{16} Joules, about equivalent to 20 megatons (20 million tons of TNT). For comparison, the bomb dropped on Hiroshima in 1945 was equivalent to 13 kilotons (13 thousand tons of TNT), more than a thousand times smaller, and equivalent to the total conversion to energy of less than 1 gram of matter.

Beyond its dire historical implications, the equation, nevertheless, hides a potentially deeper meaning. If we were, for once, to escape our own humanity and look at the more elegant side of our knowledge, we might yet progress our wisdom. Yes, we can calculate exactly what the equation means in terms of mass and destructive energy release, of course. But the question is: if we can produce so much energy from such little matter, why are we *not* doing so for peaceful and clean energy production? Indeed, the technology is on our doorstep, in the form of *fusion power* – the energy of the sun. But for the cost of a few tens of billions of dollars and the political will, humanity could have already developed this technology. The fact is that we have now reached an environmental "tipping point," and if we do not act shortly we will have reached a point of no return and a runaway greenhouse gas effect may occur. That the potential for the production of clean energy is on our doorstep and we have failed to seize it is yet one more irony in the history of human science – but it need not be so.

Underlying the energy equation $E = mc^2$ is a far deeper understanding of its significance, which will increase our knowledge tremendously. Ironically, had a scientist named Max Planck known Einstein's energy equivalence formula when he discovered his constant previously in 1900, he may well have arrived at a point where he could readily unify quantum physics and relativity.

It may be fortunate that this unification has not happened until now, but a yet more fundamental and beautiful understanding of the equation for energy exists and will be explained here. This has the potential to springboard our knowledge to the next level and perchance, with the development of new peaceful technology, will enable us to use that knowledge with prudence and wisdom.

Chapter 3

The Beginning of the Beginning

All these fifty years of conscious brooding have brought me no nearer to the answer to the question "What are light quanta"? Nowadays every Tom, Dick and Harry thinks he knows it, but he is mistaken.

Albert Einstein (1951)

It actually all began five years earlier in 1900, at a time when scientists had proudly announced, at the end of the nineteenth century, that all of physics had been discovered and that there were just a few details to tidy up. True, they had at that time made what must have seemed great advances; they had tamed the new energy form, electricity, to produce light. Also they had mastered most of mechanical engineering. But they could not know what they did not know — and what they did not know was enormous.

Around the corner lay a vast new discipline of modern Physics. What they had *not* discovered was far, far greater than what they had already discovered. In fact they had discovered very little technology — mostly because they had not mastered the principles of physics.

What was to be the beginning of the beginning of modern physics happened remarkably quickly! It was made by Max Planck, the then Dean of German Physicists working in Berlin. It happened on the 7th of October, 1900. What he discovered that evening was to be the beginning of a new revolution in physics, that which is now known as *quantum physics*, the forerunner and mainstay of modern physics.

Max Planck had in 1899 taken a relatively intense interest in a phenomenon then known as black body cavity radiation. To briefly explain, the black body equipment is a sort of blackened metallic box with a hole in it. Normally, this is heated up so that it emits infra-red radiation from the hole, which is not visible to the human eye – which is in part why we use the term black body radiation.[†] It is like heating a piece of metal, which initially gives off infra red radiation, which is not visible to the naked eye. If you give a very large amount of heat energy, it eventually becomes red hot, as red light has a greater energy than infra red. Red light is visible because it contains more energy than infra red and has a higher frequency. Similarly, as this apparatus is heated it will radiate more infra red energy and this will be of higher frequencies at higher temperatures. This suggests that the more heat energy the box appears to contain, then the higher the frequency of the energy. This would make sense, but what didn't make sense was the pattern of this radiation. There was a law called Wien's Law, which governed this, but recent experiments had disagreed with the law.

On that Sunday of the 7th of October, 1900, Max Planck was visited by the experimentalist Heinrich Rubens who had got some new data, which showed that Wien's Law was out. That evening Planck mused

---

† If you heat up the inside of the cavity enough, the emerging radiation will appear not black but dull red, then red, then orange, then yellow.

over it and developed an equation, which later then led directly to a new fundamental energy equivalence law for electromagnetic radiation.

That new equation was $E=hf$.[†] This equation looks straight forward, doesn't it? But within it resides the incredible hidden beauty of the Universe. It was the first ever equation expressing the *quantum* energy of a system. Planck perhaps did not realise the great significance of this equation at the time, and ironically its true significance is not even fully understood today. But this was the beginning of quantum physics and a new epoch in science.

Just twelve days later Planck publicly presented his data to an under-whelmed audience of his peers, who were unaware of the significance of his work. He subsequently published his seminal work later that same year in 1900[6] and an update in the following year 1901.[7] The determination of a fundamental quantum of energy, Planck's constant, was to prove critical in the development of quantum physics.

Planck initially interpreted his equation to mean that it was a mathematical subdivision of the total energy of all the atoms in the cavity, and that it was about the special case of the energy of the atoms in a cavity when they interact with light.

It was up to Einstein to later interpret, again in 1905, a meaning closer to the truth, for which he, in part, later received the Nobel Prize. It is of credit to Einstein that he was clever enough to have submitted his paper, on what was known as the photoelectric effect, to the journal called *Annalen der Physik*, where

[†] $E=hf$: the energy of a system E, is equal to a Planck's constant h, multiplied by the frequency f.

the editor of the journal was none other Max Planck himself. If he had not done this, his paper may well have been rejected.

The fortunate thing is, which Max Planck seemed to have realised, that the paper on the photoelectric effect took his equation $E=hf$ to the next level of understanding. In his paper Einstein interpreted the equation to mean that matter absorbs or emits light in individual photon packets. That is, the smallest energy quanta depended on the equation $E=hf$, such that light is not actually emitted as a continuous stream, but with the discreet energy levels $1hf$, $2hf$, $3hf$, $4hf$ and so on, with differing possible frequencies for each photon. So that, depending on the frequency f, the energy of an individual photon increases and a greater frequency results in a greater energy for each photon. So, a spectrum of radiation, such as in black body radiation will, at increasing temperatures, have individual photons at increasing frequencies.

The *true* interpretation takes us to a far more fundamental and incredibly elegant understanding of $E=hf$. Science historians have alluded that Max Planck was firmly embedded in classical physics and could not readily come to terms with what his own equation really meant. That was only initially true. Nevertheless, he had no real idea how absolutely breathtaking his new discovery really was. Indeed it seems that, to date, few have really understood the full enormously elegant potential that is held within the equation.

Even the later interpretation of the photoelectric effect, held today, falls short of the real all-embracing meaning of the equation, as Einstein was wise enough to recognise, when he said:

"All these fifty years of conscious brooding have brought me no nearer to the answer to the question "What are light quanta"?

The major problem with the current way of thinking about a photon is that each photon with a slightly different frequency would effectively constitute a different primary quantum. So, there would need to be a huge number of such different primary quanta to account for all the possible frequencies of electromagnetic radiation.

In an ultimately more sophisticated interpretation, what the equation really means is that light is quantised at a more fundamental level. Specifically, that it's components come in discreet packages, not only so far as the individual photons are concerned, *but in so as far as the frequency itself is concerned.* When considering the Planck energy itself, this comes in discreet integers, i.e. 1, 2, 3, 4, 5, 6, 7…etc. So, the energy comes in packages *not* of the discreet energy levels *1hf, 2hf, 3hf, 4hf* and so on, *but in packages of the discreet energy levels 1h, 2h, 3h, 4h, and so on.* This is an incredibly important leap in the understanding of quantum physics.

For example, each photon in the ultra violet spectrum of light has what seems like an enormous frequency of 8 hundred trillion cycles per second, or 8×10^{14} Hz (10^{14} is equivalent to 1, followed by 14 0's).

So, effectively this photon would contain 8 hundred trillion fundamental quanta. [†]

Now crucially, only one and the same fundamental quantum is overall now required. This new quantum is ephemerally small, but rather perfectly formed. For example, each photon of visible light has about a trillionth of the energy of the beat of a butterfly's wing. In comparison, the individual new fundamental quanta making up photons would have an exquisitely tiny energy of a hundredth of a trillionth of a trillionth of the energy of the beat of a butterfly's wing.

This is perhaps the ephemeral quantum Planck truly would have wanted to be able to conjure up when, in 1911, he said:

"The foundation is laid for the construction of a theory which is someday destined to permeate the swift and delicate events of the molecular world, with a new light."

The difference between the photoelectric theory and that of individual photons and this *"new light"* is enormous. It is literally as if we had discovered that the Universe were, in this given example, made of fundamental quanta one hundredth of a trillionth times smaller than previously imaginable.

In the case of a gamma ray photon, also part of the electromagnetic spectrum, the difference is even greater. For example, each photon of a gamma ray may have what is a massive frequency of ten billion, billion,

---

† n = the number of quanta *per unit time*, so it will have the same dimensions of frequency, specifically [T⁻¹]. See technical note 1 and 2.

billion cycles per second, or 10^{28} Hz (10^{28} is equivalent to 1, followed by 28 0's;). So that effectively this single photon would contain 10,000,000,000,000,000,000,000,000,000 fundamental quanta, per unit time.[†]

Again, the beauty of the conceptual difference is almost inconceivable. If Einstein had dared to think of a tiny sole photon as a single quantum then, in this case, that fundamental Universal quantum is now an exquisitely ephemeral one tenth of a billionth of a billionth of a billionth smaller than he had ever dared imagine a quantum to be. That is like comparing the volume of all the water in all the oceans on the Earth together, to a single tiny droplet of mist. [‡]

Even today the equation $E=hf$ is not fully comprehended by physics, as Einstein himself was wise enough to admit. The most common misconception is not to realise the quantised nature of frequency itself. For instance, some think you can have fractions of a unit of a fundamental frequency unit. But if you do have fractions of frequency, then obviously you do not have a complete quantum theory.

In truth, this single exquisitely ephemeral quantum, elaborated here, is an entirely logical and elegant consequence of the equation $E=hf$. In fact there appears to be no other logical way of thinking about a photon, because otherwise each photon with a slightly different frequency would indeed constitute a different primary quantum. There would, strictly speaking, need

---

[†] As the Planck energy, h, is given in *energy multiplied by seconds*, the frequency must also be *per second* (see technical note 1 and 2).
[‡] See technical note 3.

to be at least a million, trillion, trillion, trillion, different primary quanta to account for all the photon frequencies of just the electromagnetic spectrum alone. As Johann Wolgang von Goethe once said:

"Everything has been thought of before, but the problem is to think of it again".

Suffice it to say, it is important to revisit old assumptions from time to time, and the equation $E=hf$, whilst currently accepted as being the beginnings of quantum physics, does have this far more fundamental, deeper and more elegant meaning in terms of quantum physics itself.

Indeed, the equation Planck discovered that evening was only one of the two most important energy equations in scientific history. The other famous equation being $E=mc^2$, known as the Sextant equation, which was not published by Einstein until 1905, again in the same journal called *Annalen der Physik*, where Max Planck was the editor. In a second irony, had Planck known and truly understood the importance of the Sextant equation earlier, he might have been in a position to make many more intellectual quantum leaps, ones which we will show would have been able to advance science much further than it is even today. We will additionally show that it leads us directly to a more fundamental energy equivalence equation, one that unifies and beautifully explains both $E=hf$ and $E=mc^2$.

Ironically, a hundred years later, scientists again proudly announced that they were within reach of a new way of looking at the Universe, which could potentially explain everything, a theory known as string theory. True, we have made what must seem like enormous strides in the past 100 years. We have tamed space exploration, created supersonic jet aircraft and made computers, but what we are about to discover is much more far-reaching – and awesomely beautiful!

Chapter 4

Wave Particle Duality: Yin and Yang

In formal logic, a contradiction is the signal of defeat: but in the evolution of real knowledge it marks the first step in progress to victory.

Alfred North Whitehead.

Max Planck's discovery in 1900 and Einstein's later interpretation of the photoelectric effect in 1905 led to a particular contradiction. Is a photon a wave or is it a particle? Newton first described it as particle in his corpuscular theory of light. Much later on James Clerk Maxwell reverted to describing it as a wave. Einstein had gone some way to resolving the issue in his picture of the photoelectric effect. Even then he hedged his bets because of the general difficulty in accepting quantum physics at the time. Specifically, he stated that when radiation interacts with matter, energy is transferred to the electrons of the constituent atoms, "*as if*" in discrete energy quanta. So, who was right? In the final analysis, as in a lot of cases in scientific debate, both sides of the argument are ultimately correct.

However, until now, the argument has still remained unresolved. In quantum physics one simply accepts that if you use experimental devices that measure wavelike properties, you see a wave and if you use experimental devices that measure particle-like properties, you see a particle. The question "Why?" is not often even asked anymore.

Since its discovery, the concept of wave particle duality has increasingly occupied quantum physicists. The amount of research, both theoretical and experimental, conducted to study it is monumental. Moreover, the numbers of different equations covering quantum physics are becoming ever more numerous, complicated and are beginning to appear contrived. So much so that in quantum physics the term "quantum cookbook" is sometimes applied to the plethora of equations. The field of quantum physics has become so complex and counter-intuitive that it prompted one famous physicist, Richard Feynman, to say:

"I think I can safely say that nobody understands quantum mechanics".

In contrast, what we have described here, in the establishment of a fundamental quantum (see Chapter 2), not only elegantly explains the meaning of the equation $E = hf$, but aesthetically demonstrates wave particle duality from first principles.

Thus, what we find here is that the new paradigm of quantum physics leads directly to an absolutely logical explanation of wave particle duality. Given that a single photon is what acts as the single particle, then the number of fundamental oscillating quanta within that photon determines its frequency, giving it its wavelike properties. Both are integral to any photon. So, the energy quanta are arranged logically such that $E = 1h$ or $2h$, or $3h$, or $4h$, and so on.

What this means for a single photon is that the number of constituent quanta is huge. In the case of a gamma ray photon it is unbelievably large. For example each photon of certain gamma rays has what is a massive frequency of ten billion, billion, billion cycles per second, or 10^{28} Hz (10^{28} is equivalent to 1, followed by 28 0's). Effectively, this single photon would contain 10,000,000,000,000,000,000,000,000,000 fundamental quanta. [†]

Again, the conceptual difference is almost inconceivable. In this case, we see that fundamental Universal quantum is now an exquisitely ephemeral one tenth of a billionth of a billionth of a billionth smaller than a gamma ray photon. That is like comparing the volume of all the water in all the oceans on the Earth together to a single tiny droplet of mist.

With this new concept of the ephemeral universal quantum, we can now readily begin to understand quantum physics. The quintessential paradigm shift has been made. We can now express wave particle duality in one aesthetic equation.

Here it is: the next piece of magic. For as $E= hf$ and n is the number of those fundamental quanta contained within a photon, then the equation for the frequency f is:

$$f = n \qquad\qquad (1)$$

Q.E.D. Specifically, this means that frequency of a photon is equivalent to none other than the *number of*

[†] n = the number of quanta *per unit time*, so it will have the same dimensions of frequency, specifically [T^{-1}]. See technical note 2.

fundamental quanta, n_q, contained within it, per unit time †
This essentially means that whilst a photon is a discrete entity, it is made up of fundamental ephemeral oscillating quanta, which bestow upon it its frequency.

But wait; if this *were* the fundamental quantum of the Universe, one would expect matter to display some of the same type of characteristics as that of light. Indeed, it was later realised that light and matter behaved very similarly in this respect, but it was not until 1923 that the further application of the equation *E=hf* was made to matter by a young scientist called Prince Louis de Broglie.

Moreover, the equations governing the frequency and wavelength of matter would be the same for both. It is no surprise to us, with the knowledge of hindsight that this should be so; but of all the aspects of quantum physics, this was perhaps the most intriguing. That matter also does display wave particle duality has nevertheless subsequently been experimentally proven.

The details of the discovery of the wave particle duality of matter will be explained in subsequent chapters, suffice it to say that in this period of scientific history, few of these various discoveries appeared to happen sequentially. For this reason the truer and more subtle and connected meaning behind the equations have, until now, not been fully elucidated. Needless to say, the same principles that apply to wave particle duality in the photon apply to matter, and from this basis, we arrive directly at the equations of modern quantum electrodynamics.

---

† As the Planck energy, *h*, is given in *energy multiplied by seconds*, the frequency must also be *per second* (see technical note 1).

Quantum electrodynamics is now open to a full understanding. After many thousands of papers and hundreds of hefty tomes have been written about the subject of wave particle duality, we have finally arrived at a more logical and aesthetic equation for frequency. Specifically, $f = n$, such that the frequency is equivalent directly and logically to the number of fundamental ephemeral quanta contained within a system. The equation is more logical and carries more understanding, but nevertheless agrees entirely and explains both the original $E = hf$ equation *and* wave particle duality in a single elegant masterstroke.

Chapter 5

The Relativity Revolution

After a time of decay comes the turning point. The powerful light that has been banished returns.

I Ching

About the same time that Einstein was working on the photoelectric effect and his famous energy equation in 1905, he was also working on another branch of modern physics – relativity. The theory of relativity arose at a time when everybody was comfortable with the concept of Newtonian space and time. After all, Newton's theories had reigned supreme for more than 200 years and offered a degree of clockwork certainty. Points in space were fixed and events could be plotted on the fixed background of this space and time – all that was about to change radically.

Einstein again had the foresight to submit his paper, on what was known as special relativity, to the journal called *Annalen der Physik,* where the editor of the journal was none other Max Planck. Indeed, Planck's support was *as* crucial here as it had been with the publication of Einstein's earlier paper on the photoelectric effect. Almost certainly, Einstein's papers on relativity may not have seen the light of day; no other physicist at the time would have accepted these papers. Indeed, Planck was much chided for publishing such "unnecessary" work for at least a few years afterwards – until the work later came to be famous.

If it was not for these events, the name "Einstein" would probably be unknown today. Einstein wasn't even working in physics at the time; he was a grade three clerk at the Swiss patent office, and he may have otherwise sunk into obscurity. Nevertheless, the paper on relativity *was* published and the history of science was changed.

Perhaps, in the first instance, we should describe how the introduction of relativity and this forceful upheaval came about. It had been assumed in the middle of the 19th century that light travelled in space through a substance known as the "luminiferous ether". If that were the case, it was reasoned that the earth rotating on its axis and around the sun throughout the seasons would change its orientation to this ether and thus alter the measured speed of light. Historically, the most important measurement of this speed was made by Michelson and Morley in 1887. Interestingly, they found that there was *no* change in the apparent speed of light, whatever the orientation. The experiment was repeated again and again by them and others – still there was no change in the speed of light. Either there was no ether or something else very strange was going on!

In 1895, H.A. Lorentz was the first to come up with a clever solution. If the apparatus arm was going the direction the ether was contracted, then that might exactly compensate for and account for the apparent lack of difference in the speed of light. He had even developed a rather elegant equation for this contraction to provide the exact amount of contraction necessary, (see Box 1, equation 1). But, Lorentz had later to introduce a speed to a point in empty space to make the equations work when an object was moving. This is where the difficulties started, as this method seemed slightly contrived, in order to salvage the static luminiferous ether.

This is where Einstein stepped into scientific history with his theory of special relativity. Special relativity was only one of a string of Einstein's extraordinary papers to be published in 1905. This was his *annus mirabilis*, his miracle year, and the year 2005 is the centennial celebration, "The Year of Physics," in honour of that same year one hundred years ago.

Einstein changed our concept of space and time, forever. In order to explain the results of his experiment, what had pertained before regarding space and time suddenly had to disappear. The genius of special relativity was that space and time themselves were not fixed; indeed, they became almost seamlessly fluidic in nature as space-time together. Depending on the speed an object was travelling, space would shrink and time would expand. So the faster an object travelled, the shorter it became in length and the longer it would take for time to pass for that object. Specifically, clocks would actually run slower, if carried along with a moving object. This was Einstein's special relativity.

The effects regarding time and space have been proven over and over in experiments. In more modern times the effect on time has been confirmed using atomic clocks, which measure time exceedingly accurately. This effect on time has come to be known as time dilation. Equally well, the effects on length and, in particular, the increase in mass has been confirmed using particle accelerators, which can accelerate a particle to more than 99.999 % of the speed of light. The experiments confirm with great accuracy the equations, which are termed the Lorentz-Einstein transformations (see Box 1). The term "Lorentz invariance" is still used today for some of the effects of special relativity, after Lorentz, who was the first to formulate this equation:

Box 1
Lorentz-Einstein Transformations

1. $l = l_0(1 - v^2/c^2)^{1/2}$

2. $t = t_0/(1 - v^2/c^2)^{1/2}$

3. $m = m_0/(1 - v^2/c^2)^{1/2}$

where $m_0,$ $t_0,$ $l_0,$ is rest mass, time and length respectively; c is the speed of light.

Einstein, in his new paper, based his special relativity on two principles: firstly, that the speed of light was a constant for all those who observed it, whatever speed they were going and secondly, on the principle that all observers were equal. He surmised that, as the constancy of the speed of light was proven, it was sufficient to apply these rules in order to justify relativity. That is, in order to explain relativity then time would slow, and space would shrink in such a way that the speed of light would be kept constant to all observers, irrespective of their speed.

What is not entirely clear is whether Einstein wanted, at this stage, to banish the ether completely or just to modify the concept of the ether. Certainly, it is generally assumed that his 1905 paper had banished the ether. However, if we examine the actual 1905 text carefully, it suggests otherwise. The actual words that Einstein used in the introduction to his 1905 paper on special relativity[9] were:

"The introduction of a "luminiferous ether" will prove to be superfluous inasmuch as the view here to be developed will not require "an absolutely stationary space" provided with special properties nor assign a velocity-vector to a point of the empty space in which electromagnetic processes take place."

The phrase: *will not require "an absolutely stationary space"*, suggests that he was merely objecting to the presence of a static ether. An ether in some form of motion, perhaps like the air moves around with the Earth, might explain the results. An ether with no particular direction of motion, or perhaps some very fast motion of the ether would account for the results, as any motion of the Earth would then be comparatively very small and therefore would not appear to affect the results.

Today, there have been a few big surprises, particularly in Cosmology, which may rekindle the need for an ether. Although, to avoid going back to the term "ether", it has been called something else; some call it the cosmological constant, which is the term Einstein himself coined, perhaps after he realised he could not reintroduce the term ether. Some call it quintessence. Some call it space-time foam, and some scientists contend that this energy is the equivalent to the "virtual" photons that have been proposed to surround ordinary atoms to account for some of their quantum effects. This is known as "vacuum energy". But if this vacuum energy were the same as the cosmological energy, then this result would be too big. Not just a by factor of 10 times out or 1,000 times out, it would be an unbelievable factor of 123 orders of magnitude out (1 followed by 123 0's out), the biggest error margin in the history of science!

To avoid this, some call it the three-dimensional space-time lattice. But, most commonly, scientists refer to the energy inherent in space as "dark energy". Here, we will introduce the term "space-time matrix" and we will call the constituents of that matrix "quintessence". We prefer the term quintessence as it is in accordance with Aristotle's original ethereal concept of the fifth essence, which he believed was the fundamental basis for the other four essences: earth, fire, wind and water. The term has also been more recently re-introduced by the eminent physicist Lawrence Krauss, as a solution to the missing energy in the Universe.

Whatever it is called, somewhat intriguingly, the Universe is not only expanding but is doing so at an increasing rate, which implies that there is energy contained in empty space. Most scientists now accept that this something they also refer to as "dark energy" is needed to explain this mysterious phenomenon. This "dark energy" or quintessence appears to be inherent in the fabric of the space–time matrix. The old ether was banished, that is, until recently, when scientists were forced to reconsider something along the lines of the new ether (although it is now called something else).

There was another factor that could not be answered simply by the two relativity principles of the constancy of the speed of light and that all observers are equal, which Einstein used. These principles could not explain how the mass of an object increases with velocity. These principles might increase the density by shortening the length, but surely the mass itself would not change solely as a result of these observer principles. And yet, intriguingly, the mass of an object would and does also increase with increasing velocity. (see Box 1). So, there must be a reason for this mass increase, and indeed some sort of energy in empty space could account for the mass increase more logically. This could occur either by increased friction with space-time or by the accrual of energy from space-time (the actual mechanism will be discussed later). The important issue is that there is a potential reason for this mass increase and it is all linked to modern Cosmology.

If there was a Cosmological constant, however, or something equivalent to it, like a space-time matrix, there would still be a problem: what is the speed of the individual components of that matrix?

The real answer regarding the speed of a space-time matrix is quite aesthetically astonishing. Indeed, the results of relativity provide the very answer needed. The aesthetic answer is that the individual fundamental ephemeral constituents of the space-time matrix, quintessence, are themselves travelling at the speed of light. The question, what velocity would the individual essences of the space-time matrix need to have to remain constant, is answered; specifically, it is the speed of light, which is constant – so, so elegant.

The solution is thus special relativity and the individual quanta of quintessence, as they are travelling at the speed of light, will have the same velocity whatever the velocity of the object moving through them. So the solution to special relativity is to look even deeper into special relativity. We will revisit this more logical and aesthetic notion of the motion of quintessence, and this deeper meaning in the coming chapters of this book. The solution nevertheless turns out to have exquisite symmetry.

The second question is: what are these quintessence quanta? Firstly, these would need to be ephemeral in order to compose the space-time matrix, a substance that cannot be directly detected even today. Secondly, they would to have the velocity of the speed of light, and, thirdly, they would need to have some inherent energy.

The most beautiful solution is that the space-time matrix, which is composed of quintessence, are one and the same as the exquisitely ephemeral quanta that make up the photon, already described in Chapters 2 and 3. That is the ephemeral quantum that is so small that it has a tenth of a thousandth of a trillionth of a trillionth of the energy of a single photon. So there would be 10^{28} quanta (1 followed by 28 zero's) contained within a single (high energy gamma ray) photon of light. So, if a single (gamma ray) photon were equivalent the volume of all the water in all the oceans on the Earth together, then this ephemeral quantum would be equivalent to a single tiny droplet of mist.

The space-time matrix would be composed of energy of these exquisitely small quanta, according to the equation $E = hf$; so the energy comes in packages of the discreet energy levels of $1h$, $2h$, $3h$, $4h$, and so on. Effectively, the space-time matrix would be in packages of only 1 or perhaps 2, 3, 4 or several units of these exquisitely ephemeral units These would form a virtually seamless three-dimensional space-time lattice as they interweave at the speed of light to form the very fabric of space-time in a breathtakingly aesthetic way.[†]

It is with the presence of the ephemeral quantum described in Chapters 2 and 3 that we will show that the energy in empty space and, in turn, the Universe as a whole can be more logically and beautifully explained. We will show that this quintessential energy relates to none other than Planck's constant, and that the energy contained in space-time matrix conforms to the equation $E = hf$. Indeed, not only is light governed by the equation, but scientists now know matter is also governed by the very same equation. Hence, all the varied constituents of the Universe are governed by the very same equation $E = hf$. Hence, by all logical reasoning and aesthetic design, we will show an incredibly elegant solution: everything in the material Universe is composed of the same ephemeral quintessential quantum. The powerful concept of unity has effectively returned, if only by another name.

There is now a new paradigm shift; Einstein's relativity showed that space and time were interlinked

---

[†] As the Planck energy h, is given in *energy multiplied by seconds*, the frequency must also be *per second* (please see technical note1).

to from space-time. In this new paradigm energy is linked to space-time, so that we now have energy-space-time. This allows a tremendously harmonious approach to the entirety of physics.

The key to this unified quantum approach is to find a discrete quantum of mass, which would be equivalent to the quantum of energy, which Planck had found. For it to account for space-time, this quantum mass would need to be as ephemeral as the quantum unit of energy described here. It is the very presence of this quantum mass that will guide us to an understanding of special relativity, gravitation and quantum physics as a truly unified and beautifully graceful Universal design.

Chapter 6

The Battle of the Physics Giants

Discussion is an exchange of knowledge; argument the exchange of ignorance.

Robert Quillen

It was in 1913 that the next quantum leap in the understanding of physics came. The physicist Niels Bohr published three papers on the quantised nature of the atom. This was to revolutionize the view of the atom entirely. Scientists of the time viewed atoms as they would view the solar system. There was a central nucleus with tiny particles called electrons orbiting it, like the planets orbit the sun. Nowadays, we are familiar with electrons as the particles that carry current in electric wires and other electrical devices. In those times however, not much was known about electrons.

Niels Bohr discovered that these little electrons delicately and swiftly orbited the nucleus of an atom at discrete energy levels. Moreover, these particles behaved like a wave! Modern physics was already reeling from the discovery of relativity. What they had to cope with next was the fact that the components of an atom not only came in discrete energy levels, but that matter also could behave as a wave. This became known as wave-particle duality and this mysterious dichotomy has remained unresolved in modern physics, until now (see Chapter 3).

More ironic is that *had* Einstein busied himself with this subject, he could have come up with far more direct equations for quantum physics than the quantum physicists were able to come up with at the time. In truth, Einstein and the quantum physicist Niels Bohr became scientific competitors. Einstein continued to invent thought experiments that showed that quantum physics could not be right, and Bohr would write back with very clever answers that confirmed it was right. Even so, Einstein constantly questioned quantum physics, both privately and sometimes publicly. His major objection to quantum physics was the apparent probabilistic nature of quantum physics. In this regard, Einstein famously once said of quantum physics.

"God does not play dice"

In truth, the main reason why science at the time had to use the probabilistic approach was because they had not yet discovered how ephemerally tiny the fundamental quantum of the Universe really was. All they knew was the fact that Planck's constant kept cropping up. Notwithstanding this, as result of the conflict, *even* Einstein was not fully allowed in to the quantum physicists "club".

The fact was the genius equations that Niels Bohr first arrived at in 1913, for which he was later awarded the Nobel prize, were later shown to be approximations. Nevertheless, these quantised equations were the first evidence that matter such as the electron could also behave as a wave. In the Bohr atom the electrons circle the nucleus of the atom in an orbital way – but not in the classical orbital way. Each electron followed a quantised route dependant on the electron's energy. This was the origin of Bohr's quantised energies.

Ironically, even at this stage had relativistic equations been applied, as we will show, they would have been more accurate than what was devised in 1913. The fact is that at the time no one had a clue what the actual velocity of the electron was. All they knew was that it was significantly slower than the speed of light, so they could not use relativity. It was not until later in the late 1920's (see Box 2) that an Austrian physicist called Erwin Schrödinger devised a more accurate quantum equation, which now forms one of many equations in the quantum "cookbook".

Box 2
Electron Binding Energy Levels

$$En = \frac{m_e\, e^4}{8h^2\varepsilon_0^2 n^2} \approx 13.6044 \; eV/n^2$$

where m_e is the rest mass of the electron, e^4 the charge of the electron to the fourth power, h is Planck's constant, ε_0 is the permittivity of free space, n the orbital number and eV is the energy in electron volts.

We *now* know the speed of the electron, and this is such an important number in quantum physics that it has been given its own symbol, alpha (a). Indeed, the relativistic approach does work using the term a as the known velocity of the electron, as we will demonstrate later (see Box 4).

The truth is that the equation in Box 2 can also be much simplified, it turns out that, according to a standard theory called "virial theory", the binding energy of the electron in each orbit is equivalent to its classical kinetic energy ($\frac{1}{2}mv^2$, see Box 3). Specifically, the energy, which holds the electron in its orbit around the nucleus, is equivalent to the energy of its own motion, known as the kinetic energy.

In scientific terms, it turns out that it would be entirely appropriate to apply the kinetic energy formula and that the kinetic energy formula in Box 3 is equivalent to the formula in Box 2.

Box 3

Electron Binding Energy

$$En = \frac{m_e \, e^4}{8h^2\varepsilon_0^2 n^2}$$

$$a = \frac{e^2}{2\,\varepsilon_0 hc}$$

thus

$$e^4 = a^2 \, 4 \, \varepsilon_0^2 h^2 c^2$$

substituting e^4

$$En = \frac{m_e \, a^2 c^2}{2} \cdot \frac{1}{n^2}$$

as $a^2 c^2 = v^2$

$$En = \frac{1}{2} m_e v^2 \cdot \frac{1}{n^2}$$

> where m_e is the rest mass of the electron, e^4 the charge of the electron to the fourth power, h is Planck's constant, ε_0 is the permittivity of free space, n the orbital number

However, in scientific terms, it turns out that Einstein's relativistic equation for kinetic energy is known to be more accurate than using the classical formula for the kinetic energy. This is particularly true in the range of very high velocities.

The approximate equation used for classical kinetic energy, as we know from relativity, is in fact exactly that, only a good approximation. The fact remains that whilst the classical and relativistic equations agree very well at low velocities, at higher velocities the relativistic equations are more accurate. So, if we use the relativistic approach we should actually get a better answer.

The relativistic approach is straightforward, the kinetic energy is merely the total energy of the particle in motion, minus the energy that it would have if it were at rest. Since we *now* know the actual baseline velocity of the electron with a good degree of accuracy (a is about $1/137$, or 137th that of the speed of light) we can now do this calculation using the accurate value of the velocity of the electron (see Box 4).

Box 4
Relativistic Electron Binding Energy

$$En = \frac{\gamma m_e c^2 - m_e c^2}{n^2} \approx 13.6054 \ eV/n^2$$

where m_e is the rest mass of the electron, $\gamma = 1/(1 - v^2/c^2)^{1/2}$, n the orbital number and eV is the energy in electron volts.

It is clear that the relativistic equations could have been applied – and they would have given the right answers, even at this early stage of knowledge about the atom in 1913. The rather intriguing thing is that this quantum equation (Box 2) gives virtually the same answers as the relativistic approach. The truth is that relativity does not appear to have been seriously applied to this problem, because at the time they had no knowledge of the electron's actual velocity. After all, once quantum physicists had a very good equation, there appeared no reason to test others.

So, the relativistic equation gives almost exactly the same (if not more accurate) answer. Intriguingly, not only was Einstein inadvertently sidelined as far as an entry into the field of quantum physics is concerned, but if relativity had been given a chance in this aspect of quantum physics, we could now be looking at a unified approach, which would allow us to gain a truly formidable understanding of the subatomic world.

If the entry of relativity into quantum physics had occurred, at this stage, it would be far more likely that Einstein may have realised in 1913, or soon after, that his equations could account for this very quantum energy phenomenon. It was not until much later that a partially complete relativistic solution to quantum electro dynamics arose – and Einstein was not even involved. Had Einstein been involved at this stage, it is without doubt that he would have looked into the quantised aspect of relativity theory.

The key to the quantum approach was to find a discrete quantum of mass, which would be equivalent to the quantum of energy, which Planck had found. Had Einstein done this, he would have discovered the marvellously elegant and beautiful solution to what relativity and quantum physics really are.

Chapter 7

Quintessential Mass Quanta

The hypothesis of quanta will never vanish from the world.

Max Planck (1911)

Planck, shortly after his genius discovery in 1900, set about trying to determine a set of fundamental units for space, time and mass. He presupposed, quite correctly, that there should be fundamental quantities of these, which were Universal in nature and based on the constants of nature. Using the Gravitational constant and the velocity of light, he constructed these units also based on his own constant. He originally postulated a universal mass, the Planck mass.

This mass has today become enormously important, not only because this mass is postulated as the fundamental string mass in what is known as "string theory", but as this mass has also been suggested as the basis for quantum gravity.[10-16] String theory is a vastly complex field, but it has, in recent years, promised to unite all the aspects of nature. String theory is the basis for modern physics theorists saying that they are on the verge of discovering a "theory of everything" or TOE. Certainly, if this is the case then they have discovered a very big TOE.

The important question is, "is the Planck mass the right mass?" If it is not, it will be very difficult to specify a single form of string theory and very difficult to construct a theory of quantum gravity. The real problem with this standard Planck mass is that it appears to specify an "effective" maximum mass quantum and not a minimum quantum mass, which is principally required in quantum physics. The Planck mass is about the mass of a tiny grain of sand, equivalent to about 10 millionths of a gram (10µg). Although this may seem small, it is large compared to the subatomic world.

The standard Planck mass is ten billion trillion times heavier than the smallest known measured particle, known as the electron. That is equivalent to twenty-two orders of magnitude (1 followed by 22 zero's) *heavier* than the electron. This makes it awkward to use in forming quantum physics equations, which describe how the electron gracefully orbits the nucleus of an atom. Tellingly, the size of the Planck mass, when used in string theory, appears to modify the equations for quantum mechanics.[17, 18] This seems strange as these equations are based on Planck's original constant. This suggests, while Planck's constant is correct, the subsequently derived Planck mass may not be.

While the Planck energy, time and length represent a minimum quantity, the Planck mass seems to set an effective upper limit to a mass quantum. So the mass does not dovetail in with the other parameters, particularly with the Planck energy. There was no doubt about the intellectual mountain that Planck had already climbed. By all accounts, he was correct in setting his Planck units, but the problem with the Planck mass was that it appears to be an effective maximum mass. The irony here is that had Planck known the other energy equivalence formula $E = mc^2$ at the time he derived his Universal quantities, he probably would have arrived at the correct "minimum" Planck mass. This, as it turns out, solves all the difficulties with string theory.

We use the information that Planck did not have when he derived his original mass. That is, that the standard energy equivalence formula $E = mc^2$ is correct. From this we can derive a truly fundamental quintessential mass quantum (see Box 5).

Box 5
Quintessential Mass Quantum (m_q)

As $E = mc^2$

$m = E/c^2$

substitute
E for h,
then $$m_q = h/c^2 \qquad\qquad (2)$$

where h is Planck's constant, c is the speed of light. For dimensions please see technical note 2.

This suddenly changes everything; everything can now dovetail together precisely as it should. We can use this mass as the fundamental basis of mass itself. Not only for matter, but also as a component of the forces of nature such as electromagnetic energy, which is known to have what is called a non-rest mass, (that is, the apparent mass that a photon has when it is going at the speed of light). As far as the formula

$E=mc^2$ is concerned, fundamentally the mass quantum m_q, is now equivalent to the energy quantum h.

The successful paradigm for quantum physics has, to date, been to reduce the size of the constituents of the universe. As we get smaller and smaller we get closer and closer to a unified picture of how everything works. As we cut finer and finer, with sharper and sharper instruments, we learn more and more about the truer structure of the Universe. As Eden Phillpotts, a British author, poet and dramatist said:

"The Universe is full of magical things, patiently waiting for our wits to grow sharper."

In this way we have continued to reduce the size of the quintessential mass of the Universe to that which appears to be a quintessential minimum.

The question you may be asking is, how small is this new ephemeral minimal quantum mass, compared to the original Planck mass? The answer is almost impossibly small. It is a million trillion, trillion, trillion times smaller, that is 1 followed by 42 zero's smaller than the original Planck mass.

Compare the mass of a grain of sand to the mass of the Earth; you would be no-where near the smallness of this quintessential quantum compared to the original Planck mass. Expand the mass of the Earth to the mass of the solar system and compare it to the mass of a grain of sand – still not close. Expand the mass of the solar system to the mass of a thousand solar systems – still not there. Compare the mass of a tiny grain of sand to the mass of million solar systems –

that is how small the new quintessential quantum mass is compared to the original Planck mass.

We can also look at it from the other angle. Take a grain of sand and shrink it down to a mass that makes the original grain of sand equivalent to the mass of a million solar systems – that is how infinitesimally small our new ephemeral quantum mass is compared to the original Planck mass. It is vanishingly small.

It was possible even in Planck's day to have come to the same conclusions. Ironically, there *was* yet another way of proving what this minimum mass quantum should be. This might have been more apparent to Planck, had he had Einstein's energy formula to corroborate it. As previously stated, Planck had the foresight to realise that there may be a fundamental quantity of time as well as mass. In actual fact, in the original Planck's constant, energy and time are actually stitched together. If one looks at the equation ($E = hf$), then it is Planck's constant *multiplied by* the frequency that gives the total energy E of the system. As frequency is related to time, so Planck's constant has a time element. By the same token, we should do the same to derive the minimum quantum mass. Mass and time should be stitched together in the same way, to give the true quantum mass.[†] Then this mass would be equivalent to the minimum Planck's constant and give a minimum quantum mass. Indeed if we multiply the conventional Planck mass by the conventional Planck time we also get exactly the same answer for a quintessential mass quantum (see Box 6).

This quintessential mass is now, by all accounts, entirely consistent with the concept of a minimum

[†] For dimensions, please see technical note 2.

energy component, which is the fundamental theoretical and experimental basis for the Planck energy h, used in conventional quantum physics.

Box 6
Quintessential Mass Quanta (m_q)

m_q = Planck mass x Planck time

$$m_q = \sqrt{(hc/G)} \; x \; \sqrt{(hG/c^5)} = h/c^2 \qquad (2)$$

where $\sqrt{}$ is the square root, h is Planck's constant, c is the speed of light and G the gravitational constant. For dimensions, please see technical note 2.

This quantum mass now has the full nature of a fundamental quantum of mass. In keeping with the energy quantum, which comes in packages of the discreet energy levels $1h$, $2h$, $3h$, $4h$, and so on. Mass now comes in discreet packages of mass levels $1m_q$, $2m_q$, $3m_q$, $4m_q$, and so on, which in physics terms exactly match those of the energy quantum. So, the number of energy quanta is the same as the number of mass quanta.[†]

Moreover, because these mass quanta are so ephemerally small they can be used in the relativistic equations, without in any way altering the meaning or sense of these equations. Hence, we will later show that we can amplify these ideas to go on to elegantly

---

† For dimensions, please see technical note 2.

explain quantum physics and relativity itself, thereby aesthetically unifying physics.

Having done this, we can make giant leaps in the understanding of quantum physics. In the first place, we can again resolve that thorny old problem of wave particle duality, that strange quality about matter and light that gives it those properties simultaneously in quantum physics of being capable of being both a wave and a particle, with an effective mass.

We can also proceed, using these same fundamental tenets, to derive all the major formulae for quantum electrodynamics from first principles. At the end of this book, we will proceed to derive an energy equivalence equation, even more fundamental than $E=hf$, and the famous $E=mc^2$. It will then be clear that these observations are in agreement with and allow an incredibly elegant approach that unifies physics as *"one stupendous whole."*

Chapter 8

Light and Matter Waves

We are to admit no more causes of natural things than such are both true and sufficient to explain their appearances. Therefore to the same natural effects we must, as far as possible, assign the same causes.

Isaac Newton.

In the intervening years since 1900 after Plank discovered the equation $E=hf$, much of physics had been uncovered. Einstein had discovered the photoelectric effect, penned his special theory of relativity and discovered the famous equation $E = mc^2$, all in the year 1905. He subsequently developed what was then a very advanced theory of gravity, called general relativity, in 1916. We will be able to return to these theories and aesthetically unify them within quantum physics once we elaborate a better understanding of the nature of matter.

It was Niels Bohr, in his work on the Bohr atom in 1913, who gave the first hint that matter could also behave like a wave. However, it was not until 1923, ten years after that, that it was truly recognised that matter also behaved as a wave. In this case, the further application of the equation $E=hf$ was made – and it also related to matter itself. Of all the aspects of quantum physics, this was perhaps the most intriguing. That matter also displayed wave particle duality, but only if you looked at it at a small enough scale, has nevertheless been manifestly proven.

Thus, it was in 1923 that young physicist Prince Louis de Broglie hit upon the fact that matter also followed the equation $E=hf$.[19] In brief, he postulated that if light had a frequency and in turn a wavelength, then so could matter. Moreover, the equations governing the frequency and wavelength of matter would be exactly the same for both. He later won the Nobel Prize for his work. The equation he found for the wavelength of matter was dependant on h, where h was again none other than Planck's constant. The equation had earlier been shown to be the correct one for light, both experimentally and theoretically. So, the frequency and wavelength for *both* light and matter is the same and dependant upon Planck's constant. Strangely, this had been previously predicted more than 300 hundred years ago by Isaac Newton. It would seem as if Newton had almost prescient insight when, he had said:

"Are not gross bodies and light convertible into one another, and may not bodies receive much of their activity from the particle of light, which enter their composition."

Indeed, it would seem from the second part of the sentence that he had pretty much predicted the photoelectric effect. But in the first part of the sentence, Newton hints that light and matter could convert in to each other and thus they would at some fundamental levels behave in similar way.

It was not for a few centuries, that the mathematical proof would come for this. Nevertheless, the mathematical proof that de Broglie gave for deriving the wavelength of matter did come. It was, however, quite complex. It involved developing an equation for the frequency of matter, which is by no means straightforward or immediately obvious (see Box 7).

Box 7
De Broglie Frequency

$$f = \frac{m_0 c^2}{h\,(1 - v^2/c^2)^{1/2}}$$

where m_0 is the rest mass, h is Planck's constant, c is the speed of light.

Additionally, his equation involved developing the notion of what was called the "phase wave velocity". This is the concept that the speed of constituents of a wave can be different to the speed of the wave itself. As an example, the maximum velocity of sound waves in air is about 761mph (the sound barrier at sea level), but the individual air molecules are moving approximately 2 times faster than this. This notion had been previously known, but what de Broglie did was calculate the velocity of this phase wave for matter. Strangely, the phase wave velocity actually turned out to be faster than the speed of light itself. The velocity de Broglie calculated for the phase wave of matter almost always gave a velocity greater than light (see Box 8).

Box 8
Phase Wave Velocity

$$v_w = c^2/v$$

where v_w is the phase wave velocity, c is the velocity of light and v the velocity if the wave.

De Broglie was a physics PhD student in France when he proposed his idea for matter waves. The problem was that he had to postulate that the constituents of a matter wave could go faster than light. But, as we know, Einstein maintained that no physical effect could travel faster than light. In an odd twist of fate, this new idea was presented to Einstein through de Broglie's tutor. Despite, or even because of, Einstein's belief that nothing could travel faster than light and his misgivings about quantum physics, at that point in time, he held the power to alter science history.

Perhaps this is where de Broglie had been a little clever, for as part of the equation for frequency he had used Einstein's relativistic equation (see Box 7). Maybe because of this, Einstein's choice was to be open-minded and he endorsed de Broglie's theorem, to the benefit of the field of quantum physics. This acceptance was ostensibly because physicists did not count the velocity of the "phase wave" as transmitting any direct physical effect. Hence it did not, apparently, break the cosmic speed limit, the speed of light. The important thing was that the concept of a phase wave and the end equation itself were correct. The problem was, although de Broglie's equation for the actual frequency of matter was right, it was complex and conveyed no understanding. It was the actual concept behind the formula for the frequency of matter that was, in hindsight, not expressed at the most fundamental level.

Ironically, what was again lost was the opportunity to fully understand the nature of frequency, both for light and in this case for matter. All we need to know is to solve the puzzle is the *a priori* (self-evident) assertion that, as far as mass is concerned, the *total mass, is equal to the quantum mass multiplied by the number of those quanta*[†] (see Box 9, Eq. 4). From here we can arrive at far simpler equation for the frequency of matter, which nevertheless agrees with and aesthetically explains it on a fundamental level (see Box 9).

Box 9
Derivation of de Broglie Frequency

$f = n$ (Eq.1)

$n = m/m_q$ (Eq.4)

$$m = \frac{m_o}{(1-v^2/c^2)^{1/2}}$$

and

$m_q = h/c^2$ (Eq. 2)

hence

$$f = \frac{m_o c^2}{h(1-v^2/c^2)^{1/2}}$$

---

† For dimensions, please see technical note 1 and 2.

The important thing here is that we can derive a complex equation using a straightforward paradigm. In fact, there is also a more fundamental way of deriving the wavelength of matter, based entirely on an elegant and a logical understanding of the frequency (see also Chapters 2 and 3). This derivation of frequency is far more straightforward than that presented by de Broglie and sheds a unique light on the fundamentals of quantum physics. Moreover, it begins to show how the whole of physics can be united in one formidably beautiful way, which surpasses any previous expectations of the unification of physics. This is not contrary to quantum physics, but it explains it entirely on a fundamental and understandable basis.

Take the fundamental concept that we described in Chapter 2 and 3. We described the frequency of both light and matter respectively as being directly equivalent to the number of quanta it contains. The equation for the frequency f is hence:

$$f = n \qquad\qquad (1)$$

Q.E.D. Specifically, this means that frequency of matter is equivalent to none other than the *number of fundamental quanta n, contained within it, per unit time*[†]. This essentially means that the constituents of matter, such as the electron, are discrete entities; but these are made up of fundamental ephemeral oscillating quanta, which bestow upon it its frequency.

[†] n = the number of quanta *per unit time*, so it will have the same dimensions of frequency, specifically $[T^{-1}]$. See technical note 1 and 2.

Once we begin to realise that matter is also made up of this very same fundamental ephemeral oscillating quanta, then the equations for the wavelength of matter and virtually all the other equations for quantum physics pop up from first principles on a formidably elegant basis.

Let us first examine the concept of wavelength using four different waves (see Diagram 1, overleaf). If all the waves are travelling at the same speed, then same total length is travelled, as in diagram 1. In the top wave, as the frequency shows 2 peaks, then that wavelength of each cycle is also going to be large. In the second wave, the frequency shows 4 peaks, and the wavelength is half as long. The third wave has 8 peaks, and the wavelength is half as long again. In the fourth wave the frequency results in 16 peaks, and the wavelength is half as long again. Hence, we start our equation with the conventional idea that the wavelength can be calculated from the velocity of the wave divided by the frequency.

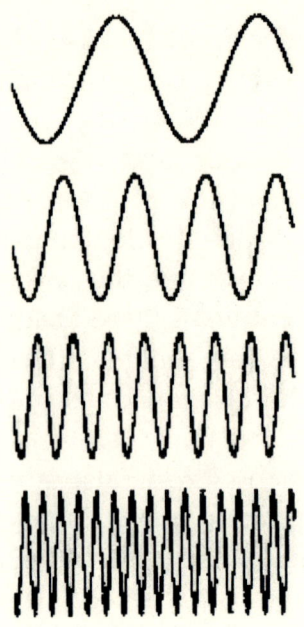

Diagram 1

Using these principles, we can readily calculate the wavelength of matter from first principles (see Box 9). Not surprisingly, we find we can get the right answer by using the much more fundamental formula for frequency. This is where we begin to show the enormous power of the fundamental quantum mass in being able to predict the equations of quantum physics. All we needed was the *a priori* (self-evident) assertion that, as far as mass is concerned, the total mass is equal to the quantum mass multiplied by the number of those quanta (see Box 10, Eq. 4).

Box 10
Matter wavelength (λ)

As:
$$\lambda = v_w/f = c^2/vf$$

$$f = n_q = m/m_q \quad (Eq.4)$$

$$\lambda = m_q c^2/mv$$

as $m_q = h/c^2 \quad (Eq, 2)$
then
$$\lambda = h/p$$

Thus, the wavelength of matter turns out to be exactly the same as light but can be derived in a much more direct way. The true meaning of this surprising finding was not understood then and is still not fully understood today. The fact that Planck's constant had once again appeared in the equations and this time, in the case of matter, had not been adequately explained. Both light and matter obeyed the equation $E = hf$, but no one knew the reason why.

The reason was quite straightforward, light and matter had been shown to behave in very much the same way as far as their energy, frequency and wavelength is concerned. So, everything points to the fact that they are made out of one and the very same thing. That is the very same exquisitely ephemeral quantum that we have described here. In this elegant model, light and matter seem to be constructed from exactly the same thing. Newton was right when he said:

"We are to admit no more causes of natural things than such are both true and sufficient to explain their appearances."

Fascinatingly, here we show that one ephemeral and elegant quantum can account for the quantum equations for both light and matter. So, if light and matter follow the same equations, then surely it might have been realised at this stage that they are composed of the same thing. Indeed, scientists have experimentally witnessed a photon turning into a particle and *visa versa* on many occasions. In truth, the problem obviously did not lay with the equation $E = hf$, but with another parameter known as the Planck mass, which Planck had originally hypothesised. Once we specify the correct Planck mass, as we do here, the equations drop out from first principles and, importantly, both equations for the wavelength light and matter agree (see Box 9).

Because the Planck mass was deduced before the equation $E = mc^2$ had been discovered, the Planck mass was not properly specified and the final link between light and matter could not be made. That is not to say that Planck was wrong; he was right but he had specified an effective maximum mass, not a minimum quantum mass. The Planck mass came out far too high and did not match the smallness of the Planck energy h.

Ironically, if Planck had got the minimum mass right, he would have been able to predict wave particle duality and in a second quantum leap maybe have gone on to develop an unified picture of how the Universe works. However, he cannot be blamed as he did not know the formula $E= mc^2$. What it probably shows is the irony of history, that once something has been set it is hard to change, and that we should have an open mind and be prepared to re-examine what we *think* we understand. Only by so doing do we progress our ideas in physics. It will be shown, here and in the next chapters, how, by using a minimum Planck mass, we arrive at a truly advanced solution.

That most advanced solution is that the constituents of electromagnetism (Chapters 2,3), space-time (Chapter 5) and now matter are made of one and the same ephemeral substance. This is an immense quantum leap in our understanding of the unity of physics.

There is far more proof to come. As all these proofs are consistent with equation $E= hf$, the energy quanta would again be arranged logically, such that $E = 1h$ or $2h$, or $3h$, or $4h$. The frequency of matter, and in turn its wave particle duality, again then emanates directly from the number of quanta it contains. The most beautiful and important part of this work is that we are able to show how light and the other forces of nature, matter and even space-time can be so elegantly constituted from exactly the same exquisitely fundamental quanta.

Chapter 9

Matter Waves Explained

"Of all the discoveries and opinions, none may have exerted a greater effect on the human spirit than the doctrine of Copernicus. The World had scarcely become known as round and complete in itself when it was asked to waive the tremendous privilege of being in the centre of the universe. Never, perhaps, was a greater demand made on mankind – for by this admission so many things vanished in mist and smoke!"

Johann Wolfgang von Goethe

It is hard to conceive the base level of understanding of science in the 1500's and the immense intellectual mountain that Copernicus had to climb. This was the beginning of the renaissance; had it not been for Copernicus, the whole of scientific history would probably have been completely different. Although his works were not widely known for nearly a century, they later became a springboard for the likes of Galileo, Kepler and Newton.

Copernicus was an unassuming Polish priest, astronomer, mathematician and also a physician. Sadly, it is said that he only received a copy of his works on his deathbed in 1543. His editor, Andreas Osiander, had, unbeknownst to Copernicus, made various modifications, which diluted the beauty and certainty of his works, to appease advocates of the geocentric theory. Nevertheless, Copernicus' work changed the way that people saw the Universe.

The strength of an observation depends on how much it is capable of explaining and how much logic and elegance it has. Galileo found Copernicus' original proposal that the earth and planets moved in orbits around the sun convincing, not because it better fit the observations of planetary motions, but because of its simplicity and beauty compared to the complicated and ever more contrived epicycles of the Ptolemaic model, which kept the Earth at the centre of the Universe.

Almost five centuries later, it may again be time to review what appears to be the ever more complex and diverse aspects of quantum physics and the sometimes conflicting fields in physics, such as relativity. Although physics has the potential to be a fascinating subject, in that it has the means to beautifully explain the workings of the entire physical Universe, it is now sadly one of the most fractionated and sub-specialised fields in science. Perhaps there is some fundamental beautiful quality that we are missing.

For instance, some fifty years ago biology was languishing in the same dilemma. Then in 1953, miraculously DNA, whose backbone was made of a simple sugar, deoxyribose, was discovered by Watson and Crick. This had the capability of explaining and unifying all of biology; after all, there was only one principle form of DNA for the entirety of all living things on Earth. Perchance, there is a simply elegant and unifying solution to the entirety of physics.

The definition of the new fundamental quintessential quantum has such enormous unifying power. Let us take that wonderfully tiny electron that swiftly and delicately orbits the nucleus of the atom. In standard quantum physics, the behaviour of this particle is governed by some rather graceful equations. But these appear to require grotesquely unwieldy mathematics to prove them. In order to understand these equations, one is asked to elaborate a view, which appears totally counter-intuitive.

For example, in order to understand an electron's orbit, we are to view the electron as "an infinitesimally small point particle with a probability density distribution". To translate this into English, we are to accept that the electron is a particle of no size at all, which has a probability of being at any place, but is in no particular place, at no particular time.

This is somewhat reminiscent of the Buddhist phrase:

"The no-mind not-thinks no-thoughts about no-things"

Buddha (563-483 B.C.)

Indeed, that is exactly how quantum physics is viewed by some physicists. Nevertheless, tempting though this no-approach is, these views obfuscate the actual absolutely true and magnificent beauty of the Universe.

To give another example of the illogical nature of quantum physics, take an electron orbiting an atomic nucleus, let's say in a page of this book; it is likely to stay orbiting that atom in this page, but there

81

remains a probability that the whole electron will find itself on the other side of the Moon or on Mars. It is ideas like this that make one think that there might be another view of the electron. Einstein found this aspect of quantum physics very difficult to accept, which is what led him to say:

"God does not play dice [with the Universe]"

Ironically, physicists may have been closer to the true origins and structure of the electron in the late 19th century. It was about this time that a physicist called Lord Kelvin was developing a theory that electrons were like smoke rings in the ether.

For years mathematicians have been modelling the appearance of the electron, and by all accounts it looks like a cloud. More recently, scientists have even directly visualised the electrons, and they found that they looked exactly like clouds, with the expected shapes.[3]

A cloud structure would also make a lot more sense in describing how an electron behaves. So, if it looks like a cloud and acts like a cloud, then perhaps it is just that, a cloud.

The question then is what is this electron cloud made of? The answer is that it is made out of the very same ephemeral quantum mass described in Chapter 7. This can be shown by mathematically predicting the correct equation for the radius of the electron cloud and this is corroborated in the next Chapter by predicting the equation, which aesthetically governs the shape of these clouds. The important thing is that this is done by using entirely logical and plausible principles.

If we dip into our quantum "cookbook" once more, we find the equation for the orbital radius of the electron (see Box 11). But this does not come readily. In actual fact, the mathematical proof for this equation spans three pages of complicated mathematical tinkering and comes with some pretty unlikely assumptions.

Box 11
Orbital Electron Radius, Hydrogen Atom

$$r = \frac{h^2 \varepsilon_0}{\pi m_e\, e^2} \approx 5.292 \times 10^{-11}\ m$$

where m_e is the rest mass of the electron, e^2 the charge of the electron to the second power, h is Planck's constant, ε_0 is the permittivity of free space, n the orbital number .

The proof for this equation is in any case not only long and complicated, but also is based on a few little approximations along the way.

The real proof for this equation can be given in three lines, not three pages (see Box 12). There are no unlikely assumptions or approximations. Indeed, all we use is the standard equation for the wavelength.

Let's start with the frequency of the electron. This can be neatly and elegantly calculated again directly from the number of quanta, which make up the electron cloud. The crucial thing here is that it is again possible to derive a very complex equation from a very straightforward paradigm. This brings a hitherto unknown degree of logic and understanding back to quantum physics.

Box 12

Orbital Electron Radius, Hydrogen Atom

$$r = \lambda/2\pi = v_w/2\pi f$$

$$v_w = c/\alpha, \quad a = \frac{e^2}{2\varepsilon_0 hc}, \quad v_{w^-} = \frac{2\varepsilon_0 hc^2}{e^2}$$

$$f = n = m_e/m_q, \quad m_q = h/c^2, \quad f = m_e c^2/h$$

hence

$$r = \frac{h^2\varepsilon_0}{\pi m_e e^2} \approx 5.292 \times 10^{-11} m$$

where m_e is the rest mass of the electron, e^2 the charge of the electron to the second power, h is Planck's constant, ε_0 is the permittivity of free space . n = the number of quanta *per unit time*, so it will have the same dimensions of frequency, specifically $[T^{-1}]$. See technical note 1 and 2.

So, the radius merely depends on the wavelength. Put quite aesthetically, the radius is dependant on the electron completing a circle with the circumference of exactly one wavelength.

As it turns out, this is an average radius, so the cloud is actually a bit spread out – pretty much exactly how you expect a cloud to be.

To get the right answer, you only have to use an entirely logical paradigm. To get an even more accurate relativistic answer, you would have to use the Einstein-Lorentz transformation equations for the mass (see Box 1). The next question is how many quanta do we need to produce this electron cloud? Well, if we take the mass of the electron (9.11×10^{-31} kg) and we divide this by the mass of a single ephemeral quantum (m_q), this gives an enormous frequency of a hundred million trillion cycles per second or 10^{20} Hz (10^{20} is equivalent to 1, followed by 20 0's). So, this electron would contain one hundred million trillion fundamental mass quanta. If we imagine the electron had the size of an average cloud in the sky and compared that to the size of a tiny droplet of mist, then this ephemeral quantum would be the same size as that droplet of mist.

The beauty is that if we determine the wavelike parameters of the electron experimentally, this is exactly the frequency that can be deduced form its wavelike properties.

This is quantum physics at its very most elegant. Yes, the electron is the smallest object we can measure the mass of, but in order to explain quantum physics we need something far, far smaller. Not only is this quantum small enough to account for the known properties of the electron, it is small enough to account for the existence of all the frequencies of the photon.

We can also, using these same fundamental tenets, derive all the major formulae for quantum electrodynamics from first principles. In the next Chapter, that beautiful equation, which describes the shapes these electron clouds take, will be derived and explained. It will then be clear that these observations are in agreement with and allow a logical understanding and an incredibly elegant unified approach to quantum physics.

Chapter 10

The Shape of Clouds to Come

"Where did we get that [equation] from? Nowhere. It is not possible to derive it from anything you know. It came out of the mind of Schrödinger."

Richard Feynman

In nature, cloud shapes in the sky may make a particular pattern depending where and how they are formed. For instance, "lee wave clouds" often form on the downwind side of a mountain. It appears that mountains often create standing waves in the atmosphere on the lee side of a mountain.

Occasionally, clouds very clearly reveal wave motion in the atmosphere. These Kelvin-Helmholz (K-H) waves are caused by differing wind speeds in adjacent levels of the atmosphere.

So it is with an electron, that delicate particle which so swiftly orbits the nucleus of the atom. Depending on its level of orbit around the atom, the electron will take a particular shape and waveform.

However, the shape of the electron cloud is probably more complex than that of an ordinary cloud. The equation is so complex that even its discoverer was honest enough to admit that he was not absolutely sure where it had come from. That physicist was called Erwin Schrödinger. Following on from the discovery of matter waves in 1923 (see Chapter 8), he was able to deduce the equation for patterns of the matter waves of the electron in 1926. His equation was revolutionary; it was literally going to change the shape of physics. But even today its origins remain obscure. A true understanding of the equation has not been reached; nevertheless, students of are largely taught that it should be used as a basis for mainstream quantum physics.

What is truly needed to formulate this equation is an understanding of how the frequency of an electron is arrived at. In the new paradigm this is intuitively dazzling (see Chapters 2and 3). Again the equation for the frequency f, is:

$$f = n \qquad\qquad (1)$$

Q.E.D. Specifically this means that frequency of an electron is equivalent to none other than the *number of fundamental quanta n, contained within it, per unit time.* [t] This essentially means that while an electron is a discrete entity, it is made up of fundamental ephemeral oscillating quanta, which bestow upon it its frequency.

Amazingly, this and the quintessential quantum mass (see Chapter 7) are all we need to proceed to derive the Schrödinger wave equation. As in Chapter 6, we know that the energy levels depend on the kinetic energy of the particular electron orbital, specifically on the energy of motion of the electron. Additionally, it was revealed in Chapter 6 that the relativistic kinetic energy using Einstein's special relativity was even more accurate than the equation for the classical kinetic

---

[t] As the Planck energy h, is given in *energy multiplied by seconds,* the frequency must also be *per second* (please see technical note 1).

energy. So, strictly speaking, we could really get a better answer using the relativistic kinetic energy. However, as the non-relativistic Schrödinger wave equation is the one we wish to demonstrate from first principles, then we will start with the classical kinetic energy to do this.

It is very difficult to actually find a short proof for Schrödinger's equation in standard textbooks. In some cases a whole textbook can be devoted to the equation and its various interpretations. One such book[5] spans 661 pages, and you still don't know where the equation comes from in the end. When you do find a proof, it usually starts with the assumptions of the de Broglie wave nature of matter, something that was never mathematically proven from first principles, until now (see Box 9). A short proof takes about a minimum of three pages of incredibly tortuous maths to come to the right equation. The logic is not always clear. Even then, scientists know its not quite right, because if you want an accurate equation for the kinetic energy at high speeds, you have to use the relativistic wave equation. In fact, a physicist named Dirac has come up with a relativistic equation, which gives a more accurate picture of the electron. Without exactly knowing it, he used a version of the relativistic equation that was demonstrated earlier (see Box 4).

In order to truly understand something and know that it is probably right, you should be able to prove something using relatively direct mathematical techniques. Here, the Schrödinger wave equation is derived using such "ordinary" mathematics in less than one page.

Box 13
The Schrödinger wave equation

$$E_K = 1/2 mv^2, \qquad\qquad E_K = \frac{m^2 v^2}{2m}$$

as: $m = m_q.n$ (eq. 4) :

$$E_K = \frac{m_q^2\, v^2\, n^2}{2m}$$

as: $m_q c^2 = h$, (eq. 2)

$$E_K = \frac{h^2}{2m} \cdot \frac{\beta^2 n^2}{c^2}$$

As: $n = f = jd\psi/dt$, and $c/\beta = dx/dt$; thus:

$$E_K = -\frac{h^2}{2m} \cdot (d\psi/dx)^2$$

as: $\psi = \sqrt{|\psi^2|}$, and $\cos x = d\sin x/dx$

$$E\psi(x) = -\frac{\hbar^2}{2m} \cdot \frac{d^2\psi(x)}{dx^2}$$

as $E\psi(x) = E_K(x) + V(x)\psi(x)$,

$$E\psi(x) = -\frac{\hbar^2}{2m} \cdot \frac{d^2\psi(x)}{dx^2} + V(x)\psi(x)$$

So, here is the proof; it is clear that the new quintessential mass (m_q) is an incredibly powerful tool with which to understand quantum physics.

90

From here we can travel light years forward in the understanding of physics. I am sure that some of our scientific readers are saying, this is all well and good, but would prefer to stick to their incredibly mystifying world of quantum physics. So, what is *new* they will ask? Well, with the new Planck mass we can go far, far further than we have gone so far, simply because all this now becomes much more comprehensible. Just by finding the right quintessential Planck mass we have come incredibly far, in the next chapter we will derive and solve some of the major problems with string theory. In the last chapter, the ultimate chapter of this book, a quantised form of the relativistic energy equation and entirely new equation for energy equivalence will be made known that is wonderfully unifying and awesomely beautiful.

Chapter 11

String Theory: Untying the Gordian Knot

"Without changing our patterns of thought, we will not be able to solve the problems that we create with our current patterns of thought."

Albert Einstein

For many centuries, the Gordian knot has represented the impossible, the intractable and often the insolvable problem. Ancient Greek legend has it that the oracle of Zeus had ordained that when it came to select a Phrygian king, the first person to ride up to the temple of Zeus in a chariot would be chosen. Gordius, father of king Midas, innocently fulfilled the oracle, as he rode his wagon to the temple and was made King. As thanksgiving, Gordius permanently tied his "chariot" to a pole in the acropolis of Gordium. The yoke was tied to a pole by an intricate knot of Cornel bark. This was so cleverly tied that the ends of the knot were hidden on the inside of the knot, and the knot seemed impossible to untie. So it stood untied for over three centuries, until the Oracle of the Delphi foretold that whomsoever could untie this knot would be the ruler of the whole of Asia.

As it happened, one day in 333 B.C., Alexander the Great visited the acropolis. At this point in time, history diverges. According to Plutarch, Alexander was unable to untie the knot and decided to *hew it asunder* with his sword. According to Aristobulus, Alexander found it easy to solve; he just knocked out the wooden dowel that held the knot to the pole and was able to slip it off and undo it from the inside. Whatever the true story, it was clear that Alexander had used a measure of lateral thinking to solve the problem. In doing so, he also fulfilled the prophecy and came to rule the whole of Asia.

In today's physics, string theory represents that Gordian knot, it is so complex that no-one can untie it.[10-12]

The principles behind string theory are worth exploring here, if only to shed light on the far more elegant truth, underneath which lies the workings of the Universe. The fundamental tenet of string theory is that the frequency of everything in the Universe is based on a vibrating string. That is, everything will vibrate in a very similar way to the frequency of the note emanating from a stringed instrument. In normal instruments, the note depends on a number of factors. The first and the most important of these factors is the mass or weight of the string itself. In string theory this mass remains the same for all strings and is given by the Planck mass (see Box 14). This is why getting the Planck mass right is so very important.

Box 14
Conventional Planck Mass (m_p)

$$m_p = \sqrt{(hc/G)}$$

where $\sqrt{}$ is the square root, h is Planck's constant, c is the speed of light and G the gravitational constant.

The second of these factors, which affects the frequency of a vibrating string, is what is known as the string tension. We have all seen orchestras tuning up their instruments and violins by turning a peg or key on the end of their instruments. By this means, they are decreasing or increasing the string tension and thus the frequency of the notes. The other factor that determines the frequency is the length of the string. Hence, the length or size of the string also determines the frequency in musical instruments.

With string theory, theorists believe they can account for the frequencies of everything: all the subatomic objects, the forces of nature and the frequency of space-time itself. So, in the case of string theory, the frequency of that wave is determined by the altering the tension of the string; the mass of the string is fixed and taken as the Planck mass, and its length is determined by measuring the size of subatomic particles. Hence, the equation for the frequency in string theory depends on much in the same way as calculating the frequency in real stringed instruments (see Box 15). In particular, everything depends on the Planck mass.

Box 15
Conventional String Frequency

$$f = \tfrac{1}{2}L \ \ \sqrt{(T/m_P)}$$

where $\sqrt{}$ is the square root, m_P is the Planck mass, L is the string length and T the string tension.

But, there are big problems with the size of the Planck mass *and* with the concept of string frequency. Because the Planck mass was so large, the corresponding string tensions had to be enormous. A subatomic particle, which weighs a miniscule amount (a thousandth of a trillionth of a trillionth of a kilogram), would need to have a (arbitrarily) large string tension of somewhere in the region of 30 tons (30,000 kg). This is very difficult to account for, and for this and other reasons, this is where string theory begins to look a little contrived.

The other major problem was that when string theory was initially devised, there needed to be ten dimensions. Normally, this is explained by an analogy to a hosepipe viewed from distance. If viewed from a distance, a hosepipe just looks like a line, but when you get up close you realise that line is actually a hose pipe, which is a three dimensional tube. This does make some sense. Particularly, if you look at the way a photon works, it travels in one direction, while having tiny vibrations in the other two space dimensions. It is these very vibrations that cause the electro and the magnetic part of the photon. So, in a solid object, there would then need to be three real dimensions. As for every real component, there would need to be an extra two vibrational components, then in a solid object there would need to be nine dimensions of space.† So, this agrees very nicely with string theory, where for every normal dimension in space there are an extra two hidden dimensions. This means there are nine space

---

† On a technical note, these nine dimensions can be returned to three by taking the square root, as in the Schrödinger wave equation (see Box 13).

dimensions and one of time, making ten. All well and good.

However, after a while, string theory became unstuck, because there were then at least 5 different mathematical solutions to the theory. What physicists did to obviate this was to simply invent another dimension, making eleven; this led to what was called membrane theory or M-theory. However, this made much less sense than the ten dimensional analogy, and this is where string theory seems to have become even more contrived to save the theory. The beauty of the approach used here is that we can *solve all the problems of string theory at a single masterstroke*, by finding what the true Planck mass is.

As we alluded in Chapter 7, whilst the Planck energy, time and length represent a minimum quantity, the Planck mass sets an apparent upper limit, to a mass quantum. So, the mass does not dovetail in with the other parameters, particularly with the Planck energy. By all accounts, Planck was correct in setting his units, but the problem with the Planck mass was that it was an apparent maximum mass. The irony here is had Planck known the other energy equivalence formula $E=mc^2$, at the time he derived his Universal quantities, he probably would have arrived at the correct minimum Planck mass. This, as it turns out, solves all the difficulties with string theory. We also arrive at a much more beautifully elegant equation for the frequency. Moreover, in this Chapter we will take string theory to the next level of understanding. We will show that we can amplify these ideas to go on to elegantly explain quantum physics and relativity itself, thereby aesthetically unifying physics.

In this unified version of string theory, using the new minimum mass, there is only one possible solution. We have described this alternative minimum Planck mass in Chapter 7 and in previous publications.[1, 2] To formulate this we used Planck's constant and the speed of light to derive a fundamental quantum mass. Earlier in this book, the validity of this fundamental mass was beautifully demonstrated, by showing that the standard quantum physical equations can be elegantly derived from first principles from it.

There are two ways in which we can derive the quintessential mass. We can use the standard energy equivalence formula $E = mc^2$. From this we can derive an exquisitely designed quintessential mass quantum (see Box 16). It is stressed that the standard energy formula may be an intermediate formula, because there is a much more exquisitely fundamental formula for energy, which is described in the next chapter.

Box 16
Quintessential Mass Quantum (m_q)

$$m_q = h/c^2 \tag{2}$$

where h is Planck's constant and c the speed of light. For dimensions, please see technical note 2.

To corroborate this, there is also a second way to derive the quintessential mass and that is to simply multiply the Planck mass by the Planck time. (see also Chapter 7).

This new perfectly designed mass quantum suddenly changes everything; everything now dovetails together exactly as it should. We can use this mass as the fundamental basis of mass itself. Not only for matter but also as a component of the forces of nature, such as electromagnetic energy, which is known to have what physicists call a non-rest mass.

Fundamentally the mass quantum m_q is now effectively equivalent in size to the energy quantum h. This quintessential mass is now, by all accounts, entirely consistent with the concept of a minimum energy component, which is the fundamental theoretical and experimental basis for the Planck energy h used in conventional quantum physics. This new fleeting quantum mass can now be used in string theory instead of the Planck mass. In keeping with the energy quantum, which comes in packages of the discreet energy levels $1h$, $2h$, $3h$, $4h$, and so on, mass now comes in discreet packages of mass levels $1m_q$, $2m_q$, $3m_q$, $4m_q$, and so on, which in physics terms exactly match those of the energy. Moreover, we can do away with everything contrived about string theory and go back to the far more elegant nine dimensional space, which is what made string theory so powerful.

Having done this, we can make massive leaps in the understanding of quantum physics. String frequency becomes graceful and subtly much more sophisticated[4]. One can get rid of those rather huge string tensions, which we mentioned earlier. In an unbelievably elegant way, the frequency of matter is then just equivalent to the number of mass quanta it contains.

As far as matter is concerned, Max Planck's genius formula $E=hf$ still holds. As a result, the previously derived equation for frequency in wave particle duality also holds (see also Chapter 2). Specifically the frequency is just equal to the number of quanta it contains

$$f = n \qquad\qquad (1)$$

Q.E.D. Specifically this means that frequency of matter is equivalent to none other than the *number of fundamental quanta n_q, contained within it, per unit time* [†]

This means we no longer need that complicated equation for the frequency, used in string theory. In the case of a photon of electromagnetism, this would be represented by an open string, like an open string of pearls. But the individual pearls or quanta would be oscillating. In the case of a particle, the string would be closed, like an oscillating fastened string of pearls. Each quantum would be ephemerally small and the string length would also be dependant on the number of qaunta contained within an object, in the same way as the frequency.

Moreover, the constituents of both matter and the photon would be the same with the same ephemeral mass quantum making up each string. The same equation for wave particle duality of matter would apply to light. So, that the equation *E=hf* would apply to matter in the same way it applies to light. Although it took until 1923 for scientists to realise this, ironically, even when they did realise it, they did not

---

[†] n = the number of quanta *per unit time*, so it will have the same dimensions of frequency, specifically [T⁻¹]. See technical note 1 and 2.

see the connection; what was needed was something that would allow the untying of the Gordian knot, and what is required is a quintessential mass.

The connection is clear, mass is also quantised – this is a crucial step forward in our understanding of the elegance of quantum physics. But the real test of this new quintessential mass quantum is just how incredibly powerful this technique is in predicting the laws of quantum physics, from first and entirely logical and aesthetic principles, which has been demonstrated in the previous Chapters.

In science things need to be provable. If all the above is not proof enough, all you need to do to prove this is to measure the length of photons of differing frequencies, then you would have your ultimate experimental proof. You would find that the length would vary directly according to the frequency, that is, according to the number of quanta.

As further proof, we will go on to derive a quantised equation for relativistic energy and also proceed to derive an energy equivalence equation even more fundamental than $E=hf$ and $E=mc^2$. It will then be clear that these observations not only corroborate the equations for quantum physics, but also point to logical understanding and an incredibly elegant unified design.

Chapter 12

Fundamental Energy Symmetry

Where order in variety we see,
And where all things differ, all agree.

Alexander Pope.

If we are to return to the history of science, by 1913 a new conflict in modern physics was already appearing. This conflict is still present, even today: relativity *versus* quantum mechanics.

In the red corner is Einstein's heavyweight equation $E=mc^2$. It is clear that in using the relativistic energy equation the quantities for energy and mass, and also length and time, were completely continuous entities. Specifically, there are no discrete stepwise values of each. In the blue corner is the quantum equation $E=hf$. In this corner the opposite is true everything appears to depend on the discontinuous or quantised unit of energy or "action" that was Planck's constant h.

That Einstein had been instrumental in establishing the validity of the quantum based equation $E= hf$ in his photoelectric theory apparently had not reconciled him, nor his peers, to the possibility that the parameters of length, time, energy and mass could or should be discrete. Nor, at any point in his distinguished career did he seriously entertain this concept or convincingly address this problem.

Equally well, for those working on quantum physics, there was no possibility of them abandoning the new concept of quantum physics. After all, why should they? More and more of their experiments on the subatomic world seemed to confirm the quantised nature of energy and matter. Even up until now, this dichotomy remains unresolved in modern physics.

The importance of Max Planck's discovery of the equation for the quantised "wave" energy in 1900 had, it seemed, taken a long time to sink in. In the meantime, Einstein had discovered his equation for "matter" energy in 1905. Soon after 1905, it was realised that the matter equation $E=mc^2$ could be applied to light waves, in a roundabout way. However, it was not till 1923 that it was found that the wave equation $E=hf$ could, in an oblique way, be applied to matter. So it would seem, from the history of science, that at least in the popular perception, the heavyweight corner $E=mc^2$ had won – if only on points.

Within the physics community, a seeming draw has been reached. When scientists know the mass m of something, they would use the equation $E=mc^2$ to calculate the energy. If they know the frequency f of something they would use the equation $E=hf$ to calculate the energy. The fact is, neither equation seems immediately obvious or logical. Why on Earth should the energy of a given piece of matter be dependant on the speed of light? Equally, why on Earth should the energy depend on frequency?

Yet again, the fundamental questions have not been asked. Why are there two equations, which seem to give the same answer for energy, and is there something more fundamental that links them both that appears more logical?

Certainly, there is something that links them, and that is the new quintessential mass described in Chapter 7. Equally well, there *is* an energy equation, which is more fundamental and, for once, entirely logical.

There is no doubt that both equations are essentially right – both have been extensively tested, and both work very well. So, it is not a question of which one is right and which one is wrong. But of which one points strongest to the more logical and unified solution. You might be surprised to find it is actually $E=hf$, for it would appear that everything in the Universe is indeed quantised at the most ephemeral level and in a most exquisite fashion.

The solution is like an enigma wrapped within a conundrum, shrouded within a mystery. It is important to use your intuitive instincts to get the correct answer but also to be guided by your logic and intellect. It is similar to a three-dimensional mathematical cryptic crossword puzzle; one needs to have sufficient clues to get the right answer, but some lateral thinking is crucial and it is vital to get some answers right before you can move to the next. Each answer interlocks with the other. Get one clue wrong and it makes it 10 times more difficult to complete. The wrong answer in this case was the Planck mass – once you have got that right, everything elegantly falls in to place (see Chapter 7).

So important is this one single fact, that the current crossword is entirely in bits, some pieces are right, some pieces are half-right. Some equations we know are pretty much right, but we have no idea where they come from. Some things we can measure accurately, but we don't know why the value is that particular value. Some parts exist that have not even been thought of yet. All of this becomes solvable and knowable. The correct Planck mass is the key to connecting two very important parts of the crossword - and filling in most of the rest.

So, here it is the *answer*, from which point everything becomes unified. In Chapters 2 and 3 we introduced the concept, where we finally arrived at a more logical and aesthetic equation for frequency. Specifically, the frequency is equivalent directly to the number of fundamental ephemeral quanta, contained within a system. Mathematically, this translated into a very straightforward equation, if we take the frequency as f and the number of quanta as n, then the equation for the frequency f, is:

$$f = n \qquad\qquad (1)$$

Q.E.D. Specifically this means that frequency of an individual system is equivalent to none other than the *number of fundamental quanta n, contained within it, per unit time* [†]

Just to recap, the difference between this concept and just viewing a photon as a single quantum is enormous. In the case of a gamma ray, each photon of

---

[†] n = the number of quanta *per unit time*, so it will have the same dimensions of frequency, specifically $[T^{-1}]$. See technical note 1 and 2.

certain gamma rays have a massive frequency of ten thousand trillion, trillion cycles per second, or 10^{28} Hz (10^{28} is equivalent to 1, followed by 28 0's). Normally, that photon would be considered as a single quantum. However, in the new model, this single photon would contain as many as 10,000,000,000,000,000,000,000,000,000 fundamental quanta.

In this case, we see that fundamental Universal quantum is now an exquisitely ephemeral one tenth of a thousandth of a trillionth of a trillionth smaller than a gamma ray photon. That is like comparing the volume of all the water in all the oceans on the Earth together to a single tiny droplet of mist.

This is the quantum leap of thought, and this is the only way it makes any sense at all. In fact, there appears to be no other logical way of thinking about a photon, because otherwise each photon with a slightly different frequency would constitute a different primary quantum. There would, strictly speaking, need to be more than a trillion trillion trillion different primary quanta to account for all of the photon frequencies of just the electromagnetic spectrum alone.

This new quintessential quantum is the fundamental energy unit of the Universe. The individual new fundamental energy quanta would have an exquisitely tiny energy of a hundredth of a trillionth of a trillionth of the energy of the beat of a butterfly's wing.

This degree of reduction in the size of energy exactly agrees with Planck's constant h, but it gives us an entirely different window on the perspective on energy. We can now reveal even more aesthetic notions. Given that $E=hf$ is correct, we arrive at an entirely logical and beautiful equation for the energy of a system, all we do is swap the f for the term n (the number of quanta), then we get the equation for the total energy of a system. So, the new equation for energy is:

$$E=hn \qquad (3)$$

Q.E.D. Put directly, this means that the energy of system is equivalent to the minimum energy quantum h, multiplied directly by the number of those quanta, per unit time.[†]

So, if your minimum energy currency is h, you directly multiply this by the number of those energy quanta, and you get the right answer for the total energy. This is just like saying, if your minimum currency is *1 cent* then you multiply it by the number of cents and that gives you the amount of currency you have in total.

This equation clearly agrees elegantly with the energy equation $E=hf$. But how does it then agree with the very famous equation $E=mc^2$? As it happens, Einstein's equation can now also be explained from first principles (see Box 17). We know what the true quantum mass is (see Chapter 7). All we need now is the *a priori* (self-evident) assertion that, as far as mass is concerned, the total mass is equal to the quantum mass multiplied by the number of those quanta (see Box 17, Eq. 4). From this, it has been shown that even the most complex quantum and energy equivalence formulae

---

[†] n = the number of quanta *per unit time*, so it will have the same dimensions of frequency, specifically [T^{-1}]. See technical note 1 and 2.

can be derived from entirely straightforward assertions. Suffice it to say that once you have seen the graceful logic of what this new concept brings in physics, it will take your breath away.

Box 17
Fundamental Energy Equation

$E = hn$

and $m = m_q n$ (Eq.4)

$E = hm/m_q$

as $m_q = h/c^2$ (Eq. 2)

$E = hmc^2/h$

and $E = mc^2$

We can go farther than this and show that the more complex relativistic energy formula also depends on a qauntised mass equation and can be directly derived from the fundamental energy equation $E = hn$ (see box 18). This is very important; because we can see at once that the fundamental energy equations are all linked, and very much depends on finding the true matter quantum. Now that we have found this quintessential quantum, we can see that relativity and quantum physics are one and the same, two aspects of the very same thing.

Using the same ephemeral quantum, we can elegantly construct everything from that same quantum. In the next book we will show how not only the electron is formed from this quantum, but also the other known fundamental particles can be formed as harmonics of the electron. Everything physical can then be derived from this exquisitely designed quantum.

Box 18
Relativistic energy Eqaution

$$E = hn \qquad\qquad (3)$$

as: $n = \dfrac{n_0}{(1 - v^2/c^2)^{1/2}}$

$$E = \frac{hn_0}{(1 - v^2/c^2)^{1/2}}$$

squaring:

$$E^2 = \frac{h^2 n_0^2}{(1 - v^2/c^2)}$$

as: $1/(1 - v^2/c^2) = 1 + \dfrac{v^2/c^2}{(1 - v^2/c^2)}$

$$E^2 = h^2 n_0^2 + \frac{h^2 (v^2/c^2)n_0^2}{(1 - v^2/c^2)}$$

as: $n^2 = \dfrac{n_0^2}{(1 - v^2/c^2)}$

$$E^2 = h^2 n_0^2 + h^2 (v^2/c^2)n^2$$

as $m = m_q n$ (eq. 4),

$$E^2 = \frac{h^2 m_0^2}{m_q^2} + \frac{h^2 (v^2/c^2)m^2}{m_q^2}$$

$$\text{and} \quad m_q = h/c^2$$

$$E^2 = m_0{}^2c^4 + v^2c^2m^2$$

$$\text{Thus} \quad E = \sqrt{[m_0{}^2c^4 + p^2c^2]}$$

Contained within this work is a very crucial paradigm shift. We have moved forward from relativity, where space and time are united to from space-time, to the unification energy with space-time, ultimately to form energy-space-time. This might have been obvious once the energy equations $E=hf$ and $E=mc^2$ were discovered, over a hundred years ago – but the quantum leap required was too great. Now that we have discovered that energy *is* inherent in space-time, the concept now becomes scientifically *de-rigueur*.

This is a giant leap forward in our unification of physics. Moreover, a new equation for energy, **$E=hn$,** emerges which is not only beautiful in its simplicity, but it shows how the two fundamental equations for energy are entirely and elegantly linked. In this sense, the new equation supersedes the other two. Primarily, it is self-evident in that it indicates the energy is dependant on the fundamental energy quantum h, multiplied by the number of those quanta. This fundamental concept is so elegant and yet so comprehensible that it removes the shrouds of mystery from the equations $E=mc^2$ *and* $E=hf$. It links these two famous equations, which means that both relativity and quantum mechanics are for once linked together, as in one harmony. This harmonious design brings an immense unity, for at once, we realise that this single new equation can bind the entire physical Universe.

End of Book 1

Technical Notes

1). *Frequency*
Common questions arise from this straightforward *a priori* assertion, $f = n$, *the frequency is equivalent to the number of quanta, per unit time*, these can readily be answered.

a.) How can a number have the dimensions of frequency? Well it is actually the number of quanta *per unit time*, so it will have the dimensions of frequency, specifically [T^{-1}].

b.) Another question is what *are* the units of time? Well, the units of Planck's constant h are given in Joule seconds (J s). Hence the unit of time of the frequency must be given in seconds (s^{-1}).

c.) A much more philosophical question arises, does it matter which units of time you use? The answer is *no*. It does not matter which unit of time you use, provided you are consistent, you get the very same answers.

This is where some people have some difficulty. The fact remains that time *elapsed* is *not* the same as *units* of time. Time can elapse, in this case, the more time that time elapses, the smaller the energy component of the minimum quantum gets as h, which consists of energy multiplied by time, is a constant. Visa versa, the less time that elapses the greater the energy component of the minimum quantum is.

Nevertheless, when we change *units*, we cannot do this in isolation, for the equation must balance. For example, if we change from S.I. units to cgs units, then not only does the meter change to centimetres, but kilograms change to grams and energy changes to ergs. To get the equivalent answer in Joules we have to convert ergs back to Joules and the same answer emerges, provided we use the same actual quantities, whatever the units. The important thing is, because we have changed one unit we also have to change other units, we cannot change units in isolation. Indeed, the equation $E=mc^2$, must hold.

This aspect is very important, so it is worth staying with the explanation. Let's now change the time unit and see what happens. The fact is, if we are using Joules to balance the equation, if we increase the time unit, we would have to increase the either length unit or the mass unit to balance the units. So, what happens when we increase the time unit? Let's say we increase the units from seconds to minutes. If we take the time elapsed for example as 1 second, then 1/60th of a minute will have elapsed, and the energy component of the quantum h will, as before, appear to rise by 60. But remembering that the length must also change means that the unit of length goes up by 60 also, as length is a component dimension of energy $[ML^2 T^{-2}]$, when the unit length component goes up the energy decreases by 60. In fact, if you change the unit of time T, you have to increase the length L dimension. The two changes balance and you get the same answer h for any new unit of time.

We can do exactly the same with time and change the unit of mass, in this case to balance the units, mass needs to go up whatever the time units went up, but squared to keep the equation balanced. It is not necessary to go through the whole explanation again to see that the two changes balance an you get the same answer h for any new unit of time.

The important thing is for every unit change the equation $E = hf$ is the same for all time units used. The main thing to remember when working this all out is to remember time *elapsed is not the same as units of time.*

This is the absolute conceptual beauty of these observations, so whatever time unit you use h is effectively the same, the frequency f is therefore the same, the number of quanta per unit time n is the same, and m_q is the same.

114

To prove this, we just need to work out for example m_q in S.I units and then in cgs and see that we get exactly the same answer.

Box 19

Lets do S.I. units first

$m_q = h/c^2$

$h = 6.626 \times 10^{-34}$ J s

$c = 2.9979 \times 10^8$ m/s

$m_q = 7.373 \times 10^{-51}$ kg s

Then lets do it in cgs

$m_q = h/c^2$

$h = 6.626 \times 10^{-27}$ erg s

$c = 2.9979 \times 10^{10}$ cm/s

$m_q = 7.373 \times 10^{-48}$ g s $= 7.373 \times 10^{-51}$ kg s

Q.E.D.

It would appear that the Universe is trying to introduce a beautiful new concept; not only is space-time interlinked, but energy and space-time are interlinked. We should have guessed this from $E=mc^2$. But now that the science is telling us that there is energy inherent in apparently empty space, that's evidence enough to support it. This takes us to the next common question.

115

2). *Dimensionalty*

The conventional formula for the Planck mass is dimensionally constrained to give a Planck mass value, with the dimensions of M, which is difficult to use in string theory.[7, 8] The quintessential mass has the dimensions [M][T], which, when multiplied by the frequency with the dimension [T^{-1}], represented by the number of quanta per unit time, we resolve the dimension back to those of M. From this result, it is also clear that, dimensionally, the number of quintessences (n) is directly equivalent to the frequency, in units of sec^{-1}. Therefore, the dimensions of the effective mass of the system, $m = m_q.n$, are entirely consistent with the dimensions of matter.

$$M = [M][T][T^{-1}]$$

These dimensions are also compatible with those of The Planck energy whose dimensions are [E][T] such that from the equation $E = hf$, we find:

$$E = [E][T][T^{-1}]$$

It is quite clear that, while the Planck energy is the key to understanding energy relations at the quantum level, it is equally important to have a fundamental mass, which conforms to the Planck scale.

3). *Volume of all the Oceans*

Volume of all the oceans = 1.37 billion km^3 =1.37 billion, billion m^3 = 1.37 billion, billion, billion mm^3. So, one tenth of a billionth of a billionth of a billionth = 0.137 mm^3 = volume of a tiny droplet of mist.

Endnotes

1 Wojciechowski, A.P. *Derivation of a Quantum Based Cosmological Energy Equivalence Formula.* www.wwk.org.uk/articles/arxiv.999.pdf.

2 Worsley, A.P. and Twist, P.J. (2001). Generation of a force on a rotating body such as a Superconductor. *Patents and Designs Journal, 5841,* 1613.

3 Polkinghorne, J.C. (1984). *The Quantum World.* Edinburgh: Longman.

4 Hay, T and Walters, P. (1987). *The Quantum Universe.* Cambridge: Cambridge University Press.

5 Byron Jr., F.W. and Fuller, R.W. (1992). *Mathematics of Classical and Quantum Physics.* New York: Dover Publications Inc.

6 Planck, M. (1900). "Zur Theorie des Gesetzes der Energieverteilung im Normalspektrum." *Verhandl. Deutsch. phys. Ges., 2,* 237.

7 Planck, M. (1901). "Über das Gesetz der Energieverteilung in Normalspektrum." *Annalen der Physik, 4,* 553.

8 Einstein A. (1905). Ist die Tragheit eines Korpers von Seinem Energiegehalt abhangig. *Annalen der Physik, 17.*

9 Einstein A. (1905). Zur Electrodynamik bewegter Korper. *Annalen der Physik, 17.*

10 Green, M.B., Schwarz, J.H. & Witten, E. (1987). *Superstring Theory*. Cambridge: Cambridge University Press.

11 Polchinski, J. (1998). *String Theory*. Cambridge: Cambridge University Press.

12 Witten, E. (1997). Duality space-time and quantum mechanics. *Physics Today, 50,* 28-33.

13 Amelino-Camelia, G. *et al.* (1998). Tests of quantum gravity from observations of γ ray bursts. *Nature, 393,* 763-765.

14 Hawking, S.W. (1978). Spacetime foam. *Nuclear Physics B, 144,* 349-362.

15 't Hooft, G. (1999). Quantum gravity as a dissipative deterministic system. *Classical and Quantum Gravity* 16, 3263-3279.

16 Wheeler, J.A. (1974). From relativity to mutability. *Revista Mexicana De Fisica, 23,* 1-57.

17 Ellis, J. *et al.* (1984). Search for violations of quantum mechanics. *Nuclear Physics B, 241,* 381-405.

18 Ellis, J. *et al.* (1992). String theory modifies quantum mechanics. *Physics Letters B, 293,* 37-48.

19 de Broglie, L. (1923). Waves and Quanta. *Comptes Rendus, 177,* 507-510.

Printed in the United Kingdom
by Lightning Source UK Ltd.
112149UKS00001BA/6